COMPUTING IN APPLIED SCIENCE

COMPUTING IN APPLIED SCIENCE

William J. Thompson

University of North Carolina, Chapel Hill

John Wiley & Sons

New York   Chichester   Brisbane   Toronto   Singapore

Library of Congress Cataloging in Publication Data:

Thompson, William J. (William Jackson), 1939 -
   Computing in applied science.

   Includes bibliographical references and index.
   1. Numerical analysis—Data processing. 2. Science—
Data processing.   I. Title.
QA297.T5   1984   519.4   83-21625
ISBN 0-471-09355-6

Printed in the United States of America

10 9 8 7 6 5 4 3 2 1

To my personal computer,
for many a fruitful evening together

PREFACE

"The purpose of computing is insight, not numbers."
R. W. Hamming, *Numerical Methods for Scientists and Engineers*, McGraw-Hill, New York, 1962.

"Their rumpled clothes, their unwashed and unshaven faces, and their uncombed hair all testify that they are oblivious to their bodies and to the world in which they move. They exist, at least when so engaged, only through and for the computers."
J. Weizenbaum, *Computer Power and Human Reason*, W. H. Freeman, San Francisco, 1976, Chapter 4, "Science and the Compulsive Programmer."

The purpose of *Computing in Applied Science* is to provide an introduction to mathematical and computational concepts and techniques used in a variety of disciplines. Among these are applied mathematics, computer science, engineering, physics, and chemistry. I have particularly emphasized understanding complex numbers and their applications, the use of finite-difference methods, power series and Fourier expansions, the formulating and solving of differential equations, and application of numerical methods in programming and data analysis.

Computer literacy is considered to be an essential part of a modern education. For students of the natural and applied sciences, it is also important to acquire computer competence. By this I mean that most applied scientists should not only be users of computer programs, but they should also have skills in formulating problems for computer solution. One of the aims of this book is to help you attain these skills. My ideas on computer competence are developed further in Chapters A1 and L1.

Who needs this book? I assume that the reader is familiar with mathematics through introductory calculus and with basic concepts from the physical sciences. The topics considered in detail are used extensively in applied mathematics, electrical and mechanical engineering, physics, and chemistry. Therefore, you will find in *Computing in Applied Science* many examples, exercises, and references from these areas of science. Table 1 presents a menu of selections for courses based on this book.

If this menu appeals to you, consider how the fare is presented.

Applicable mathematics: In the A module, "Applicable Mathematics," I develop material from mathematics, mathematical physics, and numerical analysis. In this part you will become familiar with most of the concepts and formulas required in the numerical applications in the computing laboratories module. Within each of the A chapters the relative difficulty levels of each section are indicated, using the difficulty index described at the end of this Preface. In the A chapters there are many exercises and problems to develop your analytic skills, to increase your insight, and to show the connection with other topics and with technology. The Diversion sections are designed to pro-

Table 1   Menu selections for courses based on this book.
The notation is ● for a necessary item, ⊖, for a
desirable item, and ○ for an optional item.

Chapter	Applied math	Computer science	Engineering electrical	Engineering mechanical	Physics & chemistry introductory	Physics & chemistry advanced	life science
A1	●	●	●	●	●	●	●
A2	⊖	○	●	⊖	●	⊖	○
A3	●	●	●	●	●	⊖	●
A4	●	⊖	⊖	⊖	●	⊖	●
A5	●	⊖	●	⊖	●	●	●
A6	⊖	⊖	●	●	⊖	●	●
A7	⊖	⊖	●	●	⊖	●	⊖
A8	●	⊖	○	●	⊖	●	○
L1	●	●	●	●	●	●	●
L2	⊖	⊖	●	⊖	●	⊖	⊖
L3	●	●	⊖	⊖	⊖	●	⊖
L4	●	●	●	●	●	●	●
L5	⊖	○	●	⊖	●	⊖	⊖
L6	●	●	○	○	⊖	●	⊖
L7	●	○	⊖	●	⊖	●	⊖
L8	●	⊖	⊖	⊖	●	⊖	⊖
L9	●	⊖	●	⊖	⊖	●	●
L10	●	○	●	⊖	⊖	●	⊖
L11	●	○	○	●	⊖	●	○

vide relief from the hard grind and to show the connections of the topic under study to other areas of science, technology, and endeavor.

My choice of topics in applicable mathematics has been made with the goal of providing a solid foundation in modern techniques of numerical methods and data analysis. My aim is that students should learn from *Computing in Applied Science* techniques that are of general applicability. But this learning should be through examples in which the essence of the method is not lost in a plethora of detail. For this reason, I have chosen examples that can be cast into scalar, rather than vector or matrix, form. I believe that numerical computation with matrices is better postponed until the student is very familiar with difficulties such as roundoff noise and until sufficient computational maturity has been achieved that efficient modern matrix manipulation algorithms can be understood and used.

Several of the topics in the A and L modules do not usually appear in books at this level, presumably because they are considered to be too advanced. Among these are spline fitting and the fast Fourier transform (FFT). However, these topics can be straightforwardly developed and will be widely encountered in applications. I have also tended to avoid timeworn methods when other, usually better, methods are available. For example, in the solution of differential equations Numerov's algorithm is introduced, but the Runge-Kutta algorithm is omitted.

Laboratories in computing: These form the L module, outlining the computing topics and detailing the methods for the computing exercises. The exercises have a deliberately practical emphasis on realistic examples drawn from current applications in science and technology. In particular, techniques of data analysis, such as splines, least-squares fitting, Fourier analysis and Fourier transforms, are covered in detail. Some of the laboratory chapters describe projects that are very complete, so that you will develop computer programs of realistic complexity. For example, Monte-Carlo simulations in L6 model the approach to thermal equilibrium and radioactive decay; the least-squares fitting routines in L8 can be applied to investigate the degradation of solar cells in the space environment; the fast Fourier transform program in L9 may be used to analyze brain waves; microwave resonance absorption data are analyzed in L10; and space-vehicle trajectories are obtained by numerical solution of the differential equations of motion in L11.

Many of the computing exercises suggest, but do not require, that you use programs written previously. This will accustom you to applying program modules and packages, either written by you or available through a computer system library. In particular, you will be introduced to computer graphics in an early laboratory, L4, and thereafter you can use the programs given there to display results graphically.

The main difference between the materials in the A and L modules is that in the A module general techniques of mathematics and numerical analysis are developed, whereas in the L module specific computational, numerical, and coding methods are applied to solve detailed problems. Chapters conclude with references to more-advanced numerical aspects, to scientific developments, and to extended applications of the computing methods. The 180 references are all at the current level of the reader, or slightly above. You will find few references to textbooks, but many to broad-interest science and computing magazines. I hope that you will thus widen your scientific horizons.

Programming skills needed: I assume that you are already familiar with, or are currently learning, the details of a computer language such as Pascal or Fortran. I do not try to teach you how to code a problem, but, by a few selected examples of programs, I show you what I think are examples of good programs. (The distinctions between coding, programming, and computing are emphasized in L1.) Programming of the exercises is described at three levels. First, at the mathematics and numerical analysis stage of formulas and algorithms. Second, in the form of pseudocode outlines written in an informal English-language style. The third level, reserved for programs that are of general applicability or that might be beyond the current programming experience of readers, is a program both in Pascal and in Fortran.

The fully coded programs, complete with test sections, are as follows; in L2, conversion between polar and Cartesian coordinates; in L4, graphics using printers; in L7, cubic spline fitting; in L8, straight-line least-squares fitting; and in L9 the fast Fourier transform. Each of the sample programs has been coded in Pascal and tested on an Apple II + microcomputer, using elements compatible with UCSD Pascal. These programs were also each coded in Fortran, using program statements and structures as close as possible to the Pascal version. The Fortran used is compatible with the Fortran 77 standard. Each program was tested on an Apple II + microcomputer running Apple Fortran, and it was also run without change on a macrocomputer using the WATFIV Fortran system. (Detailed specifications of the programming languages are indicated in Chapter L1.)

The program listings are in Optical Character Recognition typeface, OCR-B, so they can be directly input to a computer connected to an OCR device.

How this book is arranged: The materials in *Computing in Applied Science* may be used in many arrangements. Indeed, it is not expected that in a given course you should cover all the chapters in detail. The A and L modules are designed for concurrent use, the arrangement depending somewhat on the student's experience and on the length of the course. This book has sufficient material (including 215 exercises and 50 extensive problems) for a two-semester course, but selections from the A and L modules will make the text suitable for a one-semester course, which is one way I have been using it.

To help in the selection of course materials and to serve as a study guide a cross-reference chart for the various chapters is given after the Preface. Each section of a chapter, and each problem at the end of a chapter, has been assigned a difficulty index, using the scheme described after the Preface.

The details of program coding and testing can usually be done outside class time, which students may grow to appreciate. Many of the algorithms are also suitable for electronic (pocket) calculators, and I have indicated this when relevant. Calculators are also useful for verifying segments of programs or for making spot checks of numerical analysis aspects, such as numerical noise, discussed in A4 and L3.

Exercises and problems: Exercises within each chapter are designed so that you will develop understanding of concepts, to test whether you have really understood the derivations in the text, and to apply results from the developments in the text. The exercises in both modules will be most instructive if you work them on-line as soon as you encounter them, rather than after you have finished studying the whole chapter, because there are some vital connecting links within the exercises. The difficulty level of an exercise is about that of the section within which it appears.

Problems at the end of each chapter of the A module should be worked only after you have completed the chapter, because a problem often draws on several concepts from the current chapter, and occasionally from preceding chapters. The problems will nearly always lead you beyond the specific topics of the chapter into the real world of applied science. The difficulty of each problem, relative to others in the chapter, is indicated using the difficulty index scheme following the Preface.

If an exercise or problem is divided into sections, the divisions indicate new developments, extensions, or more demands on analytic or computing skills.

It is therefore feasible to initially work only the first sections of an exercise or problem, postponing subsequent sections until more experience has been obtained. (Students should not interpret this as a license to avoid work. See Chapter A6.2.)

The connections between, and prerequisites for, each of the A and L chapters are indicated in the succeeding cross-reference chart. This can also be a guide for instructors, or self-help students, in planning a course outline.

How this book was prepared: Production of this book was a joint effort of people and machines. A decade of students in Physics 61 used preliminary versions of the text, and J. Ross Macdonald gave many suggestions based on his teaching experience with drafts. I was helped by the editorial staff of John Wiley & Sons and by the reviewers, who will see that the worst puns have been expunged. The illustrations were skillfully prepared by Jeanne G. (Didi) Dunphey of Chapel Hill, who also helped with the book design.

The text was prepared and typeset electronically using an Apple II + microcomputer, in stand-alone mode and as a terminal to the Triangle Universities Computation Center IBM-3081 mainframe computer running the SCRIPT text-processing system (University of Waterloo, 1980) that generated code to operate a Compugraphic EditWriter 7700 photocompositor. My helpers in this were Michael K. Padrick and Robert H. Lesser of the UNC Computation Center with Roger M. Deese and staff at the UNC Printing Department. Errors in the text should be attributed to the text processors and to the author.

William J. Thompson

Chapel Hill, January 1984

HOW TO USE THIS BOOK

How to use the difficulty index: In the A module each section and each problem has a *difficulty index* associated with it, as does each section of the L module. The difficulty index has three possible values:

{1} indicates appropriate material if none of the material in the chapter has been encountered previously by the student. That is, the material is *introductory level*.

{2} is for material that is suitable if the student has good preparation for the chapter, so that the material is appropriate for *intermediate level*.

{3} indicates material of relatively *advanced level*, so that only well-prepared students should tackle it.

▶ *Exercise* 0.0.1   What is the preparation assumed of the student for each of the following:
(1) Problem [A2.4] on exponential fun.
(2) Differential equations for world-record sprints in Section A6.3.
(3) Exercise A8.5.1 on the derivation of the inverse-square force law.
(4) Testing the Nyquist criterion in a Fourier analysis, Exercise L9.3.2.
(5) Calculation of geosynchronous satellite properties, Exercise L11.1.1. □

Finding the references: The references within each chapter are given in full at the end of that chapter. The names of first authors are also given in the index, with references back to appropriate chapters.

Cross-reference chart: The following chart is a guide for selecting chapters from the A and L modules of *Computing in Applied Science*. The symbol ● indicates that the indicated chapter is *necessary* preparation for the present chapter. If ⊖ appears, then the chapter provides *desirable* preparation. The open circle symbol O indicates that the chapter is *optional* but worthwhile for preparing to read the current chapter. If there is no entry, then the two chapters are not strongly connected. For example, as preparation for the computing laboratory on least squares analysis of data (Chapter L8), you should be familiar with the material in Chapter A4 (numerical derivatives, integrals, and curve fitting); desirable preparation, but not so necessary, is found in A3 (power series expansions), L4 (introduction to computer graphics), and L7 (spline fitting and interpolation), whereas Chapter A6 (introduction to differential equations) has some material that may be useful. The other chapters are much less important for working L8.

▶ *Exercise* 0.0.2   Use the following cross-reference chart to determine with which chapters and in what depth in each chapter you should be familiar before starting to work on the following topics:
(1) Second-order differential equations (A7).
(2) Numerical approximation of derivatives (L3).
(3) Introductory computer graphics (L4).
(4) Monte-Carlo simulations (L6).
(5) Analysis of resonance line widths (L10). □

Cross-Reference Chart for Computing in Applied Science

Key: ● for necessary, ⊖ for desirable, ○ for optional.

The introductory chapters A1 and L1 have been omitted from the chart.

Topic	A2	A3	A4	A5	A6	A7	A8	L2	L3	L4	L5	L6	L7	L8	L9	L10	L11
Complex numbers & complex exponentials	A2																
Power series expansions		A3															
Numerical derivatives, integrals, curve fitting		●	A4														
Fourier expansions	●	⊖		A5													
Introduction to differential equations	⊖	●	●		A6												
Second-order differential equations	●	●	●	⊖	●	A7											
Applied vector dynamics	○	⊖	●		⊖	●	A8										
Conversion between polar & Cartesian coordinates	⊖	⊖	●					L2									
Numerical approximation of derivatives		⊖	●						L3								
An introduction to computer graphics										L4							
Electrostatic potentials by integration		⊖	●							⊖	L5						
Monte Carlo simulations			○							○	⊖	L6					
Spline fitting & interpolation		⊖	●							⊖	⊖		L7				
Least-squares analysis of data		⊖	●		○					⊖			⊖	L8			
Fourier analysis of an EEG	⊖		●	●						○	⊖	●			L9		
Analysis of resonance line widths	⊖		⊖	●	⊖	●				⊖	○			●		L10	
Space-vehicle orbits & trajectories		⊖	●		⊖	●	●	⊖	⊖	⊖							L11

CONTENTS

MODULE A   APPLICABLE MATHEMATICS †

† *Items in italics indicate extensive problems.*

MODULE L   LABORATORIES IN COMPUTING †

† *Items in italics indicate extensive programming exercises.*

COMPUTING IN APPLIED SCIENCE

A1 INTRODUCTION TO APPLICABLE MATHEMATICS

This brief chapter presents an overview of the mathematics topics that are covered in the A module, Applicable Mathematics, how these are interrelated, and how they relate to topics in the L module, Laboratories in Computing.

First, there is a section on what I mean by applicable mathematics in the context of computing in applied science. Next, motivation for using computers with this applicable mathematics is established. Finally, there are several exhortations against the abuse of computing power.

What is applicable mathematics?

For the purposes of this book, I consider applicable mathematics as math that, at the typical reader's level of mathematical and scientific experience, has direct applicability to phenomena of topical interest in science and technology.

Many of my ideas on the connections between mathematics, science, and technology are woven into the discussions appearing in the Diversion sections of various chapters. For example, the developing viewpoints on the use and interpretation of complex numbers are discussed in A2.6, the relationships between pure and applied mathematics in the last three centuries are considered in A7.2 on the history of the catenary, and the changing world views from Kepler to Bohr are emphasized in A8.4.

The applications of the mathematics of complex numbers and complex exponentials, which is developed in A2, are made in the Fourier expansions in A5 and in the fast Fourier transform algorithm derived in A5.8 and used in the EEG analysis in L9.2. Similarly, the analytical developments on power and Taylor series in A3 are used in the discussion of numerical derivatives (in A4.3) and of numerical integration (in A4.4). Taylor series expansions occur again in considering methods of numerical solution of first- and second-order differential equations in A6.6 and A7.6. The vector dynamics material in A8 is applied in the calculation of space-vehicle orbits and trajectories in L11.

All necessary results are derived within the text, with many of the steps being checked by the reader working through the exercises. The topics covered usually have much more extensive developments at a more advanced level. The References section at the end of each of the A chapters will point you to material at a level that you should be able to comprehend.

Motivation for using computers

The major reasons for using computers in numerical work are to reduce costs, relieve the boredom of human calculators, which leads to errors, and increase

1

speed. This last motivation allows computers to be used in real-time data acquisition and control.

The complexity of calculations can also be greater with computers than when human mental frailty is involved in doing the arithmetic. For examples, the use of the spline fitting methods which we derive in A4.6 and apply in L7, has developed extensively only with the growth of computer availability since the 1960s. Similarly, the fast Fourier transform (FFT), with which it is practical to compute thousands of discrete Fourier transform coefficients, has revolutionized the transformation of data. We derive an FFT algorithm in A5.8 and provide programs (in Pascal and Fortran) in L9.2, where it may be applied to the Fourier analysis of brain waves.

Computers are also useful in the numerical evaluation of integrals, as we illustrate in A4.4 and in L5 and L10, and in solving differential equations numerically, a topic investigated formally in A6.6, A7.5 and A7.6, and in L11.2 when evaluating space-vehicle trajectories. Even in simple analytical models of real-world phenomena the computer can allow rapid calculation of expressions, which permits wide variation of parameters in models. This type of application is illustrated here in the calculation of equations describing kinematics of world-record sprints (in A6.3), exponential growth during the logarithmic century (in A6.4), growth with self-constraints (the logistic-growth equation in A6.5), calculation and analysis of catenary shapes (in A7.1), modeling potentials in nerve fibers (in A7.4), and evaluation of Lorentzian resonance line shapes (in A7.7).

Against the abuse of computer power

As Weizenbaum points out in his book *Computer Power and Human Reason* (W. H. Freeman, San Francisco, 1976), it is easy to become entranced by the overt power of computers, and thereby to abuse their applications. In the applied sciences this abuse takes several forms, among which are the following:

An inappropriate model of the phenomena under study should never be employed. For example, in a straight-line least-squares analysis of data is it always appropriate to neglect the errors in the independent variable compared with those in the dependent variable? It may not be clear that the variables can be classified in this way. This question is examined, and you have a chance to investigate it, in A4.7 and L8.1, where various data, from the ages of antiquities to casualties in warfare, are given for least-squares fitting and interpretation.

The finite-arithmetic limitations of computers should not be ignored, especially because the great rate of generation of numbers by computers discourages us from checking intermediate steps in calculations. In this book we investigate numerical noise, such as round-off and truncation effects, in A4.2, in A4.3 and L3 (for numerical derivatives), and in A6.6 and A6.7 when studying the stability of differential equation solutions, especially for "stiff" differential equations. Many of the problems of arithmetic accuracy can be alleviated by choosing appropriate algorithms and carefully checking the range of applicability of the numerical methods.

Errors in programs often greatly increase these problems with using computers. I have consistently provided pseudocode outlines of the program structure and also consistently written programs for several of the exercises. I hope that you will learn by examples, rather than only by precepts.

A2 COMPLEX NUMBERS AND COMPLEX EXPONENTIALS

The motivation of this chapter is to extend the number system so that the square root of -1 has a meaning. In particular, for scientific applications the mathematical generalization allows much more powerful results to be derived than if we must avoid results in which $\sqrt{(-1)}$ appears. A number involving $\sqrt{(-1)}$ is called a *complex* number.

Historically, the first major scientific use of complex numbers was by C. P. Steinmetz (1865–1923), an electrical engineer who first used them (as we do in A2.5 and A7.4) in 1897 to simplify the analysis of alternating-current circuits.[1] In physics, chemistry, and engineering the emphasis in using complex numbers is often on the simplification of calculations. Examples of such use in this text are Fourier expansions (A5) and the solution of differential equations (A6 and A7). In quantum physics, the essence of modern natural science, the use and interpretation of complex numbers is crucial to the theory.

Our initial emphasis will be on the algebra of complex numbers. However, the analogies to such geometrical and physical quantities as two-dimensional planes, planar vectors, rotations, and phases will be developed because they greatly enrich the subject.

A2.1 {1} Definitions: The algebra of complex numbers

We define a complex number z by

$$z = x + iy \tag{A2.1}$$

where

$$i^2 = -1. \tag{A2.2}$$

Here x and y are real (conventional) numbers. The real part of z [written as Re (z)] is x, whereas the imaginary part of z [written as Im (z)] is y. If $x = 0$, then z is said to be a *purely imaginary* number; if $y = 0$, then z is said to be a *purely real* number.

The use of the term "imaginary" is historical, as we discuss in A2.6; it should not have the connotation of being unrealistic. However, its initial letter explains the use of the symbol i. For electrical engineers this i can be confused with the symbol commonly used for current, so the symbol j is sometimes used instead.

[1] It has been suggested that Steinmetz's initials be used for the unit of frequency, *cycles per second.*

The $+$ sign in equation (A2.1) denotes "associated with", in a manner to be defined, in much the same way as the $+$ sign in vector addition does not have simply the arithmetic significance. To avoid this ambiguity, we could write $z = (x,y)$, an ordered number pair. In this notation we may write similar rules to those obtained in the following, without the symbol i appearing.

Rules for complex numbers

The rules for manipulating complex numbers must be consistent with those for purely real numbers and for purely imaginary numbers. In the following let $z = x + iy$, $z_1 = x_1 + iy_1$ and $z_2 = x_2 + iy_2$. As in algebra

$$-z = (-x) + i(-y) \tag{A2.3}$$

and $z_1 = z_2$ if and only if $x_1 = x_2$ and $y_1 = y_2$.
    To add or subtract a pair of complex numbers

$$z_1 \pm z_2 = (x_1 \pm x_2) + i(y_1 \pm y_2) \tag{A2.4}$$

To multiply two complex numbers we use the expansion of brackets from algebra, to obtain after simplification

$$z_1 z_2 = (x_1 x_2 - y_1 y_2) + i(y_1 x_2 + x_1 y_2) \tag{A2.5}$$

▶ *Exercise* A2.1.1   Verify that the addition and multiplication rules, equations (A2.4) and (A2.5), hold for both purely real and for purely imaginary numbers. ☐

We define division with complex numbers to be consistent with division for real numbers and to satisfy $z(1/z) = 1$ for $z \neq 0$. Thus, if $z = x + iy$, suppose that $1/z = x' + iy'$, then $(x + iy)(x' + iy') = 1$. Using the rule for multiplication gives $xx' - yy' = 1$, $yx' + xy' = 0$. Unless both $x = 0$ and $y = 0$ (that is $z = 0$), these have the solution giving

$$1/z = x/(x^2 + y^2) - iy/(x^2 + y^2). \tag{A2.6}$$

Thus we have the division rule

$$z_1/z_2 = (x_1 x_2 + y_1 y_2)/(x_2^2 + y_2^2) + i(y_1 x_2 - x_1 y_2)/(x_2^2 + y_2^2) \tag{A2.7}$$

if $z_2 \neq 0$. If $z_2 = 0$ then the division is undefined.

▶ *Exercise* A2.1.2   Use the notation of ordered-number pairs, $z = (x,y)$ etc, to write out the preceding arithmetic rules for complex numbers. ☐

Computers and complex numbers

To appreciate why computer systems are not constructed to assume automatically that the numbers they handle are complex, consider the following exercise.

▶ *Exercise* A2.1.3   Compare the labor involved in real-number arithmetic with that for complex-number arithmetic as follows. Show that the total number of real-arithmetic operations needed for complex-number addition and subtraction is 2, for multiplication is 6, and for division is 14. (Assume that no intermediate results are stored.) □

Computer calculations using complex numbers will thus typically be about five times slower than for real numbers. If the real and imaginary parts were computed simultaneously (called *parallel computation*, which requires special computer hardware), then this time could be roughly halved. The programming examples in this book assume that real and imaginary parts of complex numbers will be computed by the long, real-arithmetic, method. For example, we suggest the following computing exercises:

▶ *Exercise* A2.1.4   Write, for computer or pocket calculator, a program that first reads X1, Y1 and X2, Y2 as the components of two complex numbers, Z1 and Z2, then writes out the results of their addition, subtraction, product, and quotient. For the quotient the program should first check that Z2 is not zero. The program should then loop to read more values. What is a good way to indicate that no more calculations are to be made? Run the program for many interesting pairs of complex numbers. □

▶ *Exercise* A2.1.5   Compare and contrast the numerical properties of complex numbers with those of vectors by considering the following:
(*a*) Given $z_1 = 3 + i4$ and $z_2 = -4 + i3$ calculate $z_1 \pm z_2$, $z_1 z_2$, z_1/z_2. The preceding exercise may be useful for this and other numerical evaluations of complex-number expressions.
(*b*) Given the two vectors with (x,y) components $(3,4)$ and $(-4,3)$, calculate their sum and difference. Show that their scalar product is zero.
(*c*) Can the product of two complex numbers be zero without at least one of them being zero? □

After some experience with the numerical properties of complex numbers, we consider some more formal properties.

▶ *Exercise* A2.1.6   To appreciate some of the formal differences between complex-number and vector arithmetic, consider the coordinates of the vectors in two dimensions (x_1,y_1) and (x_2,y_2). Show that the sign-reversal, addition, and subtraction rules are the same as for the corresponding complex numbers. Note that multiplication of complex numbers does not correspond to the scalar product of vectors. □

Complex numbers and geometry

Clues as to a geometric interpretation of complex numbers are given by the following properties of i:

$$i^2 z = -z, \tag{A2.8}$$

$$i^{-1} z = -iz, \tag{A2.9}$$

and

$$i^3 z = -iz, \tag{A2.10}$$

$$i^4 z = z. \tag{A2.11}$$

These properties of i show that when it multiplies a number z it produces a similar effect to that of a counterclockwise rotation through $\pi/2$ acting on a vector. Similarly i^{-1} corresponds to rotation through $-\pi/2$.

► *Exercise* A2.1.7   Two trigonometric formulas involving addition of angles are

$$\cos(\theta_1 + \theta_2) = \cos\theta_1 \cos\theta_2 - \sin\theta_1 \sin\theta_2, \tag{A2.12}$$

$$\sin(\theta_1 + \theta_2) = \sin\theta_1 \cos\theta_2 + \cos\theta_1 \sin\theta_2. \tag{A2.13}$$

Use these to give a geometric interpretation of the multiplication rule for complex numbers. □

► *Exercise* A2.1.8   Find a general rule for simplifying i^n where is a positive or negative integer. (Recall that the *modulo* function, mod (p,q), is the positive remainder when p is divided by q.) What is the rotation interpretation of this rule? □

We will return to the geometric interpretation of complex numbers in A2.2. In the meantime, some more notation and terminology is appropriate.

Complex conjugation, modulus, argument

Complex numbers that are related to each other by just a sign reversal of the imaginary part are frequently needed. We therefore introduce a special notation; the *complex conjugation* operation. Mathematicians usually denote it by a bar, ‾, other scientists often use an asterisk, *, preferring stars to bars. In the latter notation

$$z^* = x - iy \tag{A2.14}$$

if and only if

$$z = x + iy. \tag{A2.15}$$

Therefore, using the rules for complex numbers, we can readily derive the following properties of the complex conjugation operation: $z^* = z$ if and only if z is real,

$$z + z^* = 2 \text{ Re } (z), \tag{A2.16}$$

$$(z - z^*)/i = 2 \text{ Im } (z), \tag{A2.17}$$

$$(z_1 \pm z_2)^* = z_1^* \pm z_2^*, \tag{A2.18}$$

$$(z_1 z_2)^* = z_1^* z_2^*, \tag{A2.19}$$

$$(z_1/z_2)^* = z_1^*/z_2^*. \tag{A2.20}$$

The last three equations show that the operation of complex conjugation is distributive with respect to addition, subtraction, multiplication, and division. The relation

$$zz^* = x^2 + y^2 \tag{A2.21}$$

shows that zz^* is zero only if z is identically zero. It also provides a quick way of deriving the division formula from that for multiplication. Write z_1/z_2 as $z_1 z_2^*/(z_2 z_2^*)$, then use the multiplication rule (A2.5) to produce

$$z_1/z_2 = [(x_1 x_2 + y_1 y_2) + i(y_1 x_2 - x_1 y_2)]/(x_2^2 + y_2^2) \tag{A2.22}$$

The frequent occurrence of zz^* and its connection with vectors in two dimensions leads us to a notation, the *modulus* of a complex number z, defined by

$$\text{mod } (z) = |z| = +\sqrt{(x^2 + y^2)} = +\sqrt{(zz^*)}. \tag{A2.23}$$

This indicates the magnitude of z if we consider it as a vector lying in the x - y plane. The modulus is also called the *absolute value*. For example, the modulus of a real number is just its value without regard to sign. For the modulus of a complex number to be zero, both its real and imaginary parts must be zero.

Just as a vector can be characterized by a magnitude and an angle, it is useful to introduce an angular variable for complex numbers. We define the *argument* of z = x + iy by the requirement that

$$\cos [\text{ arg } (z)] = x/[\text{ mod } (z)], \tag{A2.24}$$

and

$$\sin [\text{ arg } (z)] = y/[\text{ mod } (z)]. \tag{A2.25}$$

These two requirements are necessary and sufficient to define arg (z) to within an integral multiple of 2π.

▶ *Exercise* A2.1.9   Show that the common formula tan $[\text{ arg } (z)]$ = x/y is not sufficient to define arg (z). □

The argument is also called the *phase angle*. The phase angle is sometimes confusingly called the *amplitude*. However, the latter term is more commonly used for the magnitude of a complex number. If the pair (x,y) formed the components of a vector, then arg (z) would be the angle that it makes with respect to the positive x axis. For example, arg $[\text{ Re } (z)]$ = $\pm\pi$, arg (i) = $\pi/2$, arg $(-i)$ = $-\pi/2$.

▶ *Exercise* A2.1.10   We derived the representation that iz means "Rotate z in the left-hand screw sense (counterclockwise) by $\pi/2$." Is this consistent with the result iz = $(-y)$ + ix? Describe the arguments and geometric arrangement of x and y to achieve this. □

We now consider some numerical examples relating to the complex conjugate, modulus, and argument of complex numbers.

▶ *Exercise* A2.1.11   Write a program (for computer or pocket calculator) that reads X,Y as the components of a complex number Z, then computes and outputs Z^2, $|Z|^2$, mod (Z), arg (Z). For the argument the program should check that $Z \neq 0$; why? The program should then loop to read another X,Y pair. What is a sensible input to terminate the program? □

▶ *Exercise* A2.1.12
(*a*) Calculate z^2, $|z|^2$, mod (z), and arg (z) for the following complex numbers z; $3 + i4$, $-4 + i3$, $i4$, -4, $i3$, $-i4$, $-i3$. The preceding exercise will be useful (if the program is correct) to accomplish this.
(*b*) Are the first two numbers at right angles? (How can *numbers* be at right angles?) □

A2.2 {1}   The complex plane: De Moivre's theorem

In A2.1 we discovered that many properties of complex numbers correspond to those of two-dimensional vectors and that i multiplying a complex number changes it in a way that corresponds to rotation of a vector by $\pi/2$. Further, the multiplication rule has a geometric interpretation. Robert Argand[1] promoted many of the ideas which led to the use of a diagram in which each pair

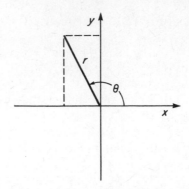

Figure A2.2.1   The complex plane (Argand diagram), showing the Cartesian coordinates and the polar coordinates of a complex number.

of real numbers (x,y) defines a point z in a plane, as shown in Figure A2.2.1. This diagram is called the *Argand diagram* or *complex plane*, designed to make complex numbers plain. The choice of axes maintains the agreement with Cartesian coordinates and also ensures that iz corresponds to a left-handed rotation of z about an axis perpendicular to the x - y plane.

The correspondence between complex numbers, plane geometry, and vectors very convenient, because understanding of one aspect may help clarify others, and results from the theory of planar vectors may sometimes be used for complex numbers and vice versa. Thus the complex plane bridges geometry and algebra. The hidden-treasure problem [A2.1] illustrates this very nicely.

The relation between rotation and multiplication with complex numbers is developed as follows.

[1]Robert Argand (1768–1822) was a Swiss who worked as a bookkeeper in Paris and emphasized the use of complex numbers. The complex-plane representation was first published by the Norwegian Caspar Wessel (1745–1818). Students in Germany learn about the "Gaussian plane," after C. F. Gauss (1777–1855). The bicentenary of his birth was commemorated in Germany by a postage stamp showing the complex plane. Science has both patriotic and international aspects.

Cartesian and plane-polar coordinates

In any plane these two coordinate systems are related as in Figure A2.2.1 by

$$x = r \cos \theta, \, y = r \sin \theta, \tag{A2.26}$$

where r is the radius and θ is the polar angle. Thus

$$z = x + iy = r[\cos \theta + i \sin \theta], \tag{A2.27}$$

with

$$r = \text{mod} \, (z) \geqslant 0 \quad \text{and} \quad \theta = \arg \, (z). \tag{A2.28}$$

The *principal value* of the polar angle θ is the smallest argument that satisfies $-\pi < \theta \leqslant \pi$. The principal value thus allows a unique specification of a point in the complex plane in terms of plane-polar coordinates.

The complex conjugate of z is readily obtained as

$$z^* = r[\cos \theta - i \sin \theta] = r[\cos \, (-\theta) + i \sin \, (-\theta)]. \tag{A2.29}$$

Thus z and z^* are related as reflections across the x axis. In the language of modern physics, a purely real number ($z^* = z$) has positive parity, because it is left unchanged by this reflection, and a purely imaginary number ($z^* = -z$) has negative parity, because it changes sign on reflection. (Purists may object to this parity analogy by arguing that it refers to a fictitious space, the complex plane, whereas parity refers to real space; but, how "real" is real space?)

▶ *Exercise* A2.2.1
(*a*) Write, for computer or pocket calculator, a program that reads R and THETA, computes the Cartesian components X and Y and outputs them, then loops for more input. What is a good way to terminate program execution?
(*b*) The polar coordinates of a set of complex numbers are (as $[r,\theta]$) $[5, -\pi/3]$, $[1,\pi/2]$, $[2, -\pi]$, $[\pi, -\pi/2]$. Find the corresponding Cartesian coordinates.
(*c*) Write, for computer or pocket calculator, a program that first inputs X,Y, then computes and outputs R and THETA for the polar coordinates r and θ. Note that an earlier programming exercise (A2.1.11) gave these as mod (z) and arg (z). Before beginning the calculation the program should check that the complex number is not zero; why? Note that θ is in radians; if you don't like this, make a program option to allow the angle to be input in degrees. The program should be able to loop to read more values. What is a good way to indicate that there are no more values?
(*d*) Given the following complex numbers, find r and θ; $3 + i4$, $-4 + i3$, i, -1, $-i$.
   The Pascal and Fortran programs presented in computing laboratory L2 provide a good start for this exercise. ☐

Now that we have an angular representation of complex numbers, we can discover several interesting results.

De Moivre's theorem

The French mathematician Abraham De Moivre (1667–1754) enunciated an interesting theorem on multiplication of complex numbers. We now derive and use this theorem. Consider the multiplication of two complex numbers expressed in polar-coordinate notation

$$z_1 = r_1[\cos \theta_1 + i \sin \theta_1] \tag{A2.30}$$

and

$$z_2 = r_2[\cos \theta_2 + i \sin \theta_2]. \tag{A2.31}$$

The product $z_1 z_2$ is simplified by using trigonometric identities (A2.12) and (A2.13). Thus we obtain straightforwardly

$$z_1 z_2 = r_1 r_2[\cos (\theta_1 + \theta_2) + i \sin (\theta_1 + \theta_2)]. \tag{A2.32}$$

Therefore, when complex numbers expressed in polar form are multiplied, their product is obtained by multiplying their moduli and adding their arguments. This addition to produce multiplication is the same as the rule for combining exponents when numbers are multiplied, and thus it provides a clue to Euler's theorem in section A2.3.

Division of complex numbers in polar coordinate representation is easily accomplished by noting that $1/z = z^*/|z|^2$, then using (A2.29) to obtain

$$1/z = [\cos (-\theta) + i \sin (-\theta)]/r \tag{A2.33}$$

The direction of $1/z$ is thus the opposite of z and the magnitude $|1/z| = 1/r$. Therefore, using the multiplication rule in polar-coordinate form, we have the rule for division of complex numbers

$$z_1/z_2 = z_1(1/z_2) = (r_1/r_2)[\cos (\theta_1 - \theta_2) + i \sin (\theta_1 - \theta_2)]. \tag{A2.34}$$

The polar-coordinate expressions for multiplication and division are certainly much easier understood (and remembered) than the corresponding expressions in Cartesian coordinates.

Exercise A2.1.3 emphasized that complex-number multiplication and division in Cartesian form are much slower than the real-number operations. We now see that polar-coordinate form, using only r and θ values, simplifies the complex-number operations. However, addition and subtraction in polar form are time-consuming because of the required conversions using trigonometric functions. This behavior is analogous to that of logarithms.

▶ *Exercise* A2.2.2   As examples of multiplication and division using polar-coordinate form, show that $iz = r(\cos \theta' + i \sin \theta')$, where $\theta' = \theta + \pi/2$, that
$i^4z = z$, and that $1/i = i^* = -i$. Interpret these results in terms of the
Argand diagram. ☐

To generalize these results for the product of two complex numbers to the
product of n such numbers, we have that

$$z_1 \ldots z_n = r_1 \ldots r_n [\cos (\theta_1 + \ldots \theta_n) + i \sin (\theta_1 + \ldots \theta_n)]. \qquad (A2.35)$$

In particular, if all the z numbers are the same, then we have *De Moivre's
theorem*

$$z^n = r^n [\cos (n\theta) + i \sin (n\theta)]. \qquad (A2.36)$$

A very direct proof of this theorem is obtained from the preceding equation.
A proof of De Moivre's theorem by the method of induction is offered as
Problem [A2.2].

A2.3 {1}   Complex exponentials: Euler's theorem

In the preceding section we noticed that the multiplication of complex numbers by addition of their arguments has much in common with the exponen-
tial function, for which, if $z_2 = \exp (a)$ and $z_2 = \exp (\beta)$ with a and β numbers, then $z_1z_2 = \exp (a) \exp (\beta) = \exp (a + \beta)$. What is an appropriate
choice of the exponents for complex numbers z?

Deriving Euler's theorem

We begin the derivation of Euler's theorem by making the abbreviation
$\cos \theta + i \sin \theta = E(\theta)$. By taking derivatives with respect to θ we see that

$$D_\theta E(\theta) = -\sin \theta + i\cos \theta = i(\cos \theta + i \sin \theta) = iE(\theta). \quad (A2.37)$$

Is there another form of $E(\theta)$? If so, it must satisfy the derivative relation just
derived. Familiarity with derivatives may inspire us to consider the exponen-
tial function $a \exp (\beta\theta)$, which satisfies $D_\theta [a \exp (\beta\theta)] = \beta[a \exp (\beta\theta)]$ for
any constants a and β. Agreement with the preceding equation is obtained
only if $\beta = i$. Further, since a is independent of θ, we can determine a by
choosing $\theta = 0$ and noting that, $a = a1 = a \exp (i0) = E(0) = 1$. Thus,
$E(\theta) = \exp (i\theta)$, and we have proved *Euler's theorem*

$$\cos (\theta) + i \sin (\theta) = \exp (i\theta). \qquad (A2.38)$$

This remarkable theorem, showing a profound connection between geometry and algebra, was enunciated in 1748 by Leonard Euler (1707–1783), a citizen of Basel in Switzerland.

Euler's theorem makes it clear why multiplication of complex numbers involves addition of angles, because the angles add when they appear in the exponents.

Formal relationships between the trigonometric and complex exponential functions can be established by noting that

$$\exp(-i\theta) = \cos(-\theta) + i\sin(-\theta) = \cos(\theta) - i\sin(\theta). \quad \text{(A2.39)}$$

By combining this with formula (A2.38) for $\exp(i\theta)$, we get

$$\cos(\theta) = [\exp(i\theta) + \exp(-i\theta)]/2 = \text{Re}[\exp(i\theta)] \quad \text{(A2.40)}$$

and

$$\sin(\theta) = [\exp(i\theta) - \exp(-i\theta)]/(2i) = \text{Im}[\exp(i\theta)]. \quad \text{(A2.41)}$$

Note that the trigonometric functions are real if θ is real.

▶ *Exercise* A2.3.1   Use the complex-exponential forms of cosine and sine to prove the well-known identity from geometry *Pythagoras' theorem*, $\cos^2\theta + \sin^2\theta = 1$, after the mathematical philosopher Pythagoras of Samos (*ca*.560-480 *B.C.*). □

Special examples of Euler's theorem are

$$\exp(i\pi/2) = i, \ \exp(i\pi) = -1, \ \exp(3i\pi/2) = -i, \ \exp(2i\pi) = 1. \quad \text{(A2.42)}$$

Thus the properties of i from section A2.2 have been rediscovered. (What is the interpretation of these results in terms of polar diagrams?)

Applying Euler's theorem

As an example of the use of Euler's theorem, we consider briefly the problem of complex roots of equations, which until now you probably rejected as meaningless. Suppose that we want the cube roots of -1, one of which is -1 itself. Since $-1 = \exp(i\pi)$, the 1/3 power of both sides gives as another cube root $\exp(i\pi/3) = \cos(\pi/3) + i\sin(\pi/3) = (1 + i\sqrt{3})/2$. The third cube root can be obtained by noting that -1 can also be reached by a rotation by π from the real axis. Thus it is just the conjugate of the other complex root.

The exponential form $\exp(i\theta)$ is mathematically much simpler than either of its components, $\cos(\theta)$ and $\sin(\theta)$. For example, $\exp[i(\theta_1 + \theta_2)] = \exp(i\theta_1)\exp(i\theta_2)$ is much simpler than the forms given by (A2.12) and (A2.13). This simplicity is exploited in the fast Fourier transform algorithm developed in A5.8 and in L9.2. Also, when taking derivatives we have that $D_\theta \exp(i\theta) = i \exp(i\theta)$, a result which is neater than $D_\theta \cos\theta = -\sin\theta$ and $D_\theta \sin\theta = \cos\theta$.

▶ *Exercise* A2.3.2   Use Euler's theorem to derive the derivative formulas for cosine and sine from that for $\exp(i\theta)$. □

A2.4 {1}   Hyperbolic functions

The study of differential equations in A6 and A7 requires exponential functions with complex exponents. A frequently occurring combination involves exponentially damped and exponentially increasing functions. Suppose that in the formulas for the cosine and sine in terms of the exponential we extend the notion of angle to include complex values. This leads to the definition of hyperbolic functions.

Definition of hyperbolic functions

We make the choice that θ is a purely imaginary number $\theta = -iu$, where u is a real number. Thus $u = i\theta$ and we define the hyperbolic functions

$$\cosh(u) = \cos(iu) = [\exp(u) + \exp(-u)]/2, \qquad (A2.43)$$

the *hyperbolic cosine (cosh)*, and

$$\sinh(u) = -i\sin(iu) = [\exp(u) - \exp(-u)]/2, \qquad (A2.44)$$

the *hyperbolic sine (sinh*, pronounced ''sinsh''). The hyperbolic functions are real-valued functions if u is real. The name hyperbolic comes from noting that by Pythagoras' theorem $\cos^2(iu) + \sin^2(iu) = 1$, which reduces to

$$\cosh^2(u) - \sinh^2(u) = 1. \qquad (A2.45)$$

As u varies, this equation defines parametrically a rectangular hyperbola $x = \cosh(u), y = \sinh(u)$, with lines at $\pi/4$ to the x and y axes as asymptotes; $x^2 - y^2 = 1$.

   Graphs of the *cosh* and *sinh* functions can be sketched by noting that for x large and positive $\cosh(x) \approx \sinh(x) \approx \exp(x)/2$, that when x is small $\cosh(x) \approx 1$, $\sinh(x) \approx x$, and that $\cosh(x)$ is symmetric about $x = 0$ whereas $\sinh(x)$ is antisymmetric about $x = 0$, as shown in Figure A2.4.1 .

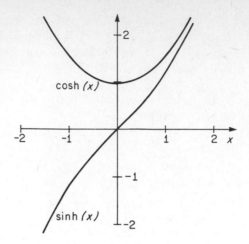

Figure A2.4.1   Hyperbolic cosine (cosh) and hyperbolic sine (sinh) functions.

▶ *Exercise* A2.4.1

(*a*) Write functions for computer or pocket calculator that will compute the hyperbolic functions cosh (x) and sinh (x). Use these functions in a program that computes them for a range of x over which the difference between the two functions exceeds 1%. The step size of x should be just small enough to allow accurate plotting.

(*b*) Make a plot, by hand, pocket calculator, or computer, of the hyperbolic functions. Do you need to calculate for arguments of both signs? Does your plot resemble Figure A2.4.1 for the hyperbolic functions? □

By comparison with the hyperbolic functions, the cosine and sine functions are called *circular functions* because they parametrically describe the unit circle centered on the origin;

$$\cos^2(\theta) + \sin^2(\theta) = 1, \qquad\qquad (A2.46)$$

with $x = \cos\theta$, $y = \sin\theta$ and $x^2 + y^2 = 1$. The different behavior of hyperbolic and circular functions is explored in Problems [A2.5] and [A2.6].

The cosine and sine functions are displayed in Figure A2.4.2. Their behavior should be contrasted with that of the hyperbolic functions, from which they differ by the interchange of real and imaginary arguments.

Circular-hyperbolic analogs

Identities for hyperbolic and circular functions are simply related by the following rule:

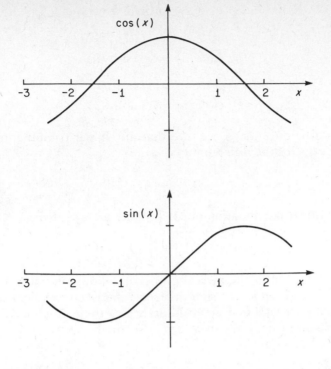

Figure A2.4.2 The circular functions, cosine (cos) and sine (sin).

An identity for hyperbolic functions is the same as that for the corresponding circular functions, except that in the former the product (or implied product) of two *sinh* functions has the opposite sign to that for two *sin* functions.

(A2.47)

► *Exercise* A2.4.2 Provide a general proof of the hyperbolic-circular identity rule (A2.47). ☐

For example;

$$\cosh(2x) = \cosh^2 x + \sinh^2 x, \tag{A2.48}$$

whereas

$$\cos(2x) = \cos^2 x - \sin^2 x. \tag{A2.49}$$

► *Exercise* A2.4.3 Prove the following hyperbolic analog of De Moivre's theorem;
$[\cosh(x) + \sinh(x)]^n = \cosh(nx) + \sinh(nx).$ ☐

The derivatives and integrals of hyperbolic functions follow from their definitions;

$$D_u \sinh(u) = -\cosh(u), \tag{A2.50}$$

$$D_u \cosh(u) = +\sinh(u). \tag{A2.51}$$

By differentiating once more, we find that any linear combination of $\cosh(x)$ and $\sinh(x)$ satisfies the differential equation

$$D_x^2 y = +y, \tag{A2.52}$$

whereas any linear combination of $\cos(x)$ and $\sin(x)$ satisfies

$$D_x^2 y = -y, \tag{A2.53}$$

When investigating the numerical properties of differential equations in A7, we will find that the innocent sign difference between the right-hand sides of these two equations makes a great difference to the ease of solving equations like these. Equation (A2.52) may give rise to the numerically troublesome "stiff" differential equation.

A2.5 {2}   Phase angles and vibrations

The angle θ in the complex plane often has an important interpretation in physical applications. For example, we use it in A7 when studying rotary, oscillatory, or vibratory motion. Suppose that the real angle $\theta = \omega t$ where ω is the constant angular velocity and t denotes time. Then

$$z(t) = \exp(i\theta) = \exp(i\omega t) \tag{A2.54}$$

describes in the complex plane uniform circular motion of the point z, while x and y describe simple harmonic motions. To prove this periodicity, note that

$$z(t + T) = \exp[i\omega(t + T)] = \exp(i\omega t)\exp(i\omega T) = z(t), \tag{A2.55}$$

wherein the last equality is satisfied if the period T is related to the angular frequency by $T = 2\pi/\omega$. By writing out the real and imaginary parts of $z(t)$ we have that

$$x(t) = \cos(\omega t), \quad y(t) = \sin(\omega t), \tag{A2.56}$$

which describe simple harmonic oscillations along the x and y axes, respectively. The phase angle between these two motions is $\pi/2$. The particular circular motion studied has unit amplitude [$r(t)$ = 1 for all t], because $|z(t)|$ = 1.

▶ *Exercise* A2.5.1   A simple-harmonic, vibratory motion is represented by $x = A \cos(\omega t) = \text{Re}(z)$, where z is the related complex number. At t = 1 sec, $z = 3 + i4$. What is the amplitude A of the vibrations? What are possible values of the angular frequency ω on the basis of the information given? Give an intuitive explanation of why there is more than one value. □

Chapter A5 on Fourier expansions and several of the computing laboratories develop the topic of expansions into harmonic motions. Extensive use of complex-exponential representations will be made and the interpretation of phase angles will be developed. Problem [A2.3] provides the first steps to construction of the fast Fourier transform algorithm.

A clear exposition of the relations between complex exponentials, vibrations, and waves is presented in the book by French.

A2.6 {2}   Diversion: The interpretation of complex numbers

In the early days of the natural sciences, and especially before the quantum physics revolution of the 1920s, it was usual for scientists to apologize for the use of complex numbers as a mathematically convenient trick; all observable quantities were to be obtained by forming purely real numbers from quantities used in calculations. For example, it was great Euler, who contributed much to pure and applied mathematics (as well as to A2.3), who coined the term "imaginarius" for $\sqrt{(-1)}$. In the book *Kandleman's Krim*, J. L. Synge gives an amusing presentation of the difficulties that scientists had in overcoming this prejudice against complex numbers. The imaginableness of imaginary numbers is also discussed in an interesting *Scientific American* article by Martin Gardner.

Nowadays, scientists are more willing to consider complex numbers as directly meaningful. (However, I never did convince a senior colleague, since retired from teaching.) Physicists (and a few engineers, and even fewer chemists) have become quite adventuresome in exploring the complex plane. Consider a scattering experiment in which data have been obtained for scattering angles in the angular range $0 \leqslant \theta \leqslant \pi$. What would be observed if the data could be extrapolated into the complex-angle plane? Such an extrapolation process is called *analytic continuation*. Similarly, measurements made at real energies or frequencies can be extrapolated to complex energies or frequencies. The methods of analytic continuation have been particularly fruitful in high-energy and nuclear physics. Problem [A2.6] provides an exercise in analytic continuation. This topic is further developed in the investigation of the Lorentzian resonance function in A7.4 and in Problem [A7.4].

The development of the interpretation of complex numbers provides an example of the consequences of education, and of the dominance of scientific thought by mathematical representation. Some scientists would claim that a phenomenon is not understood until it is expressed mathematically. For more on the relation between the "queen and servant of science" and the natural sciences, read the articles by Browder, and Wigner.

Problems with complex numbers

[A2.1]   {1}   (Finding buried treasure) In section II of his book *One, Two, Three ... Infinity*, George Gamow discusses how complex numbers may be used to solve a geometric problem involving a map and buried treasure. To paraphrase the instructions for locating this treasure: There were two trees and a gallows on a flat island. To find the treasure, walk from the gallows to the first tree; turn to the right through a right angle and walk the same distance; stake this point; from the gallows walk to the second tree; here turn left by a right angle and pace out the distance just covered; the treasure is halfway between here and the stake.

To hang it all, on reaching the island the treasure seekers could not find the gallows.

(*a*) Use the complex plane and complex numbers to locate the treasure; a convenient location for the gallows is at Γ.

(*b*) Show that the result is independent of Γ.

(*c*) Solve this problem using either Cartesian or Euclidean geometry.

[A2.2]   {2}   (Proofs of De Moivre's theorem)

(*a*) Prove De Moivre's theorem by the method of induction. (This method has very interesting relations to loop structures in computer programming.) To do this, suppose that the theorem is true for n, then show that it must hold for $(n + 1)$. Then all it needs is the result demonstrated for a particular n, for example, $n = 0$.

(*b*) The suggested method of proof holds only for n a positive integer. Does the theorem hold for all integers n?

(*c*) Use Euler's theorem to prove De Moivre's theorem for arbitrary n.

[A2.3]   {1}   (Exponentials and the FFT) In deriving the fast Fourier transform (FFT) algorithm in A5.8 the distributive and recurrence properties of the complex exponential function are vital. Let $E(N) = \exp(-2\pi i/N)$.

(*a*) Consider the properties of the powers of $E(N)$. Show that $E^0(N) = E^N(N) = 1$, $E^{-1}(N) = E^*(N)$ and also that $E^{a+b}(N) = E^a(N)E^b(N)$, $E^{pa}(N) = E^a(N/p)$ for any a,b and for $p \neq 0$.

(*b*) Hence show that if $N = 2^\nu$ with ν integral, then no matter how many even integral powers of $E(N)$ are required, they can all be calculated by evaluating the exponential function only once, then by multiplying powers of $E(N)$.

These results, and the reduction of a problem involving N Fourier coefficients into successively smaller subsets of lengths $N/2$, $N/4$, ..., 2, produces the economy of fast Fourier transform calculations.

[A2.4] {1}   (Exponential fun) The special relations between the exponential function and i are emphasized in the following, perhaps surprising, calculations:
(a) The well-known "sailor's formula" is $c = ii$. What is the numerical value of c, and what is the nationality of the sailor?
(b) Determine the numerical value of McDonald's constant $e^i e^{io}$.
(c) The temperature T of polar oceans has been expressed by the formula $T = icc$ in Celsius. Here $c = \exp(i\pi/4)$. What is the value of T?
(d) Why did Peary and Amundsen make use of polar diagrams?

[A2.5] {1}   (Graphing hyperbolic functions) Compute, using a small computer program or pocket calculator, the hyperbolic functions cosh (u) and sinh (u). For the same range of u compute cos (u) and sin (u). In all cases use the reflection symmetry (parity) properties of these functions to simplify the labor. Graph the four functions on the same scale, and discuss those properties that are important in physical applications.

[A2.6] {3}   (Analytic continuation) As an example of analytic continuation, consider the analytic properties of $E(z) = [\exp(z) + \exp(-z)]/2$ as follows:
(a) Sketch $E(z)$ *vs* $z = u$ for u real. Hopefully, you will produce the graph of the function cosh (u).
(b) On an axis at right angles to this sketch $E(z)$ *vs* $z = i\theta$, for θ real; do you get cos (θ)? Thus far $E(z)$ is easy.
(c) Now let z be an arbitrary complex number and sketch $E(z)$ *vs* z, using the same axes as in (b). This will prove difficult unless you have complex three-dimensional graph paper. A compromise for those who have only real two-dimensional graph paper is to draw contours of Re $[(E(z)]$ on one graph and contours of Im $[E(z)]$ on another graph. Can you visualize the analytic continuation of the cosine function off the real axis and into the complex plane? (This is also the analytic continuation of the cosh function off the imaginary axis.)

References on complex numbers

Browder, F. E., "Does Pure Mathematics Have a Relation to the Sciences?," in *American Scientist*, **64**, 542 (1976).
French, A. P., *Vibrations and Waves*, W. W. Norton, New York, 1971.

Gamow, G., *One, Two, Three ... Infinity*, Viking, New York, 1947.

Gardner, M., "The Imaginableness of the Imaginary Numbers," in *Scientific American*, August 1979, p.18.

Synge, J. L., *Kandleman's Krim*, Jonathan Cape, London, 1956.

Wigner, E. P., "The Unreasonable Effectiveness of Mathematics in the Natural Sciences," in *Symmetries and Reflections*, Indiana University Press, Bloomington, 1967.

A3 POWER SERIES EXPANSIONS

This chapter presents power series, with an emphasis on their practical applications and numerical properties. In particular, the Taylor series expansion is extensively studied. It forms the basis for many of the approximation schemes developed later, such as the binomial approximation in A3.5, the numerical approximation of derivatives and integrals in A4.3 and A4.4, and the numerical solution of differential equations in A6.6 and A7.6. The emphasis is frankly practical; many formal results on the algebraic properties of power series are given in Flanders *et al.*, Hansen, Jolley, and Rainville. These references have thousands of formulas for the summation of power series.

Looking ahead in this chapter, we provide in section A3.1 motivation for the extensive use of power series in applicable mathematics and discuss the convergence of series in A3.2. The Taylor series, polynomial approximations, and the binomial approximation are covered in A3.3, A3.4, and A3.5. As a diversion, we consider financial interest schemes in section A3.6. A series of problems and further references on series conclude this chapter.

A3.1 {1} Motivation for using power series

Series of positive powers of an independent variable are of great importance in applicable mathematics: First, since the algebraic properties of such series have been well studied, there are many general results available from mathematicians. Second, because the coefficients of successive powers are often straightforwardly related to each other, algorithms for generating high-order coefficients iteratively from lower-order coefficients can usually be found. Such algorithms are then very suitable for programming in loop structures, so that the numerical coefficients and the series up to a given number of terms can be evaluated automatically.

Let us define a few terms: A *series* in arithmetic is a sum of terms each of which can be obtained from the preceding terms by a general rule. An *infinite series*, often also just called a series, has an unlimited number of terms, the sum of the first n of which is called the partial sum S_n. If the value of S_n approaches a definite limit as $n \rightarrow \infty$ the series is said to *converge*; otherwise, the series is said to *diverge*. A *power series* is a series in which successive terms are related through fixed powers of some variable, say x.

Series appear in scientific applications when successive effects are combined by a simple rule. For example, a basketball rebounding to successively smaller heights travels a total vertical distance given by the series whose terms are the distances traveled between any two bounces. In such applications the scientist hopes that the series converges — if it does not, then results such as nonconservation of energy (in the example of a basketball) are predicted.

Sometimes very slow convergence of a series will lead to numerical inaccuracies (see A4.2) and long computing times, suggesting that the representation chosen is inappropriate and that the problem may have been tackled in the wrong way. The plethora of analytical results on power series, which is very helpful in formal solutions, often tempts the novice (and occasionally the experienced numerical analyst) into using them uncritically for numerical evaluations. This temptation should be avoided like strong wine.

One use of series is to simplify formulas. For example, the series $1 - 3x + 3x^2 - x^3 = (1 - x)^3$ is usually easier to interpret if the right-hand expression is used. On the other hand, noting that $(1 - x)^3$ tends to $1 - 3x$ when $|x| \ll 1$, may reduce a problem from a cubic to one approximately linear in x. Therefore we will also use power series in approximation formulas to simplify the manipulations and interpretation. In particular, Taylor's theorem in A3.3 is a very powerful general result that enables us to expand many functions in power series. Finally, we note that a polynomial in x of order p is formally a power series, the coefficients of which are zero for powers of x that are negative or that are greater than p.

A3.2 {1}   Convergence of power series

In studying numerical methods for computing in applied science we emphasize the practical problems associated with the convergence of series. Many series that can be mathematically proven to be convergent will not converge numerically in a satisfactory way because of the finite accuracy of the computing device. For example, if a large number of terms is required to get a relatively accurate sum, then errors from round-off or subtractive cancellation may produce large errors, as we discuss in A4.2. On the other hand, if the mathematics proves that the series is divergent, do not attempt to sum it. The series may appear to be converging, but sooner or later it will blow up!

One can easily be confused by verbal arguments about series, as in the paradox attributed to the ancient Greek philosopher Zeno: "I can't go from here to the wall, for to do so I must first cover half the distance, then half again what still remains, and so on; this process can always be continued and can never be completed." Do you think that Zeno would have understood the process of taking a limit and the differential calculus?

We write the partial sum of a series as

$$S_n = \Sigma \, a_k, \qquad\qquad (A3.1)$$

in which the summation over k ranges from 1 to n. The coefficients a_k are called the *terms* of the series. If S_n tends to a definite value as $n \rightarrow \infty$, then this value, say S, is called the sum of the infinite series.

The geometric series

The geometric series is a power series that can be readily investigated and that has many scientific applications. It is obtained from the formula for the terms

$$a_k = ar^{k-1}, \tag{A3.2}$$

with $a \neq 0$ and $r \neq 0$. Thus, $a_1 = a$, $a_2 = ar$, $a_3 = ar^2$, etc. In this series r is considered to be the independent variable. The geometric series is illustrated graphically in Figure A3.2.1.

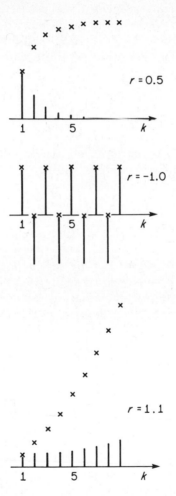

Figure A3.2.1   Terms in the geometric series are shown as solid bars, and the sum to this number of terms is shown crossed, for three multipliers: 0.3 (convergence), and − 1.0 and 1.1 (both produce divergence).

Successive terms in the geometric series can be generated by the *iteration formula*

$$a_{k+1} = ra_k, \quad a_1 = a. \tag{A3.3}$$

That is, r is the ratio of successive terms in the series. Consider the sum of the first n terms of the geometric series, S_n, given by

$$S_n = a + ar + ar^2 + \ldots + ar^{n-1}. \tag{A3.4}$$

We can investigate the convergence of the geometric series in some special cases: If $r = 1$, then $S_n = na$ and the series has no limiting value as $n \to \infty$, so the series diverges. If $r = -1$, then $S_n = 0$ for n even, but $S_n = a$ for n odd, so the partial sum does not have a definite limit as $n \to \infty$. Therefore the series does not converge.

Geometric series summed

To proceed further with geometric series, we derive a closed formula for the sum to n terms. Given the preceding partial sum, take r times it and subtract the two equalities; all the terms cancel, except the first and the last. So we obtain

$$S_n = a(1 - r^n)/(1 - r), \tag{A3.5}$$

except for $r = 1$, when we know that the series diverges anyway. Alternative proofs of this result can be obtained by using the method of induction (Problem [A3.2]) or by considering the relation between S_n and S_{n+m} (Problem [A3.4]). If $r^n \to 0$ as $n \to \infty$, then the geometric series converges to

$$S = a/(1 - r) \tag{A3.6}$$

for $|r| < 1$. Problem [A3.1] generalizes the question of convergence of the geometric series to r a complex number.

A physical example that approximates the geometric series is the bouncing of a fairly elastic ball from a hard, flat surface. Under these conditions, r, here called the "coefficient of restitution," is approximately independent of the distance of fall. (An experiment to check on this has been presented by Maurone.) We know that, from conservation of total energy, the ball cannot bounce higher on successive bounces, and that the total vertical distance through which it moves is finite, even if we ignore the fact that a real ball is of finite size and will stop bouncing once its rebound height is less than its radius.

Numerical aspects of computing and using the geometric series are given in Problem [A3.3].

▶ *Exercise* A3.2.1

(a) Show that the series $1 - (1 - \varepsilon) + (1 - \varepsilon)^2 - (1 - \varepsilon)^3 + \ldots$ for ε positive and $\leqslant 1$, converges to $1/(2 - \varepsilon)$.

(b) For $\varepsilon \to 0$ the sum tends to $\frac{1}{2}$. Give a simple explanation of this result. □

The geometric series provides an example of a series whose convergence can be investigated directly and for which a closed expression for the sum can be obtained. In the next section a more general and very powerful theorem on series expansions of functions is proven and its applications are illustrated.

A3.3 {2}   Taylor series and their interpretation

In the numerical applications in later chapters we will often require polynomial approximations to functions. Brook Taylor (1685–1731), an English mathematician, proved a very powerful theorem that can be used with continuous, very smooth, and well-tailored functions.

Taylor's theorem states that if a function $f(x)$ and its first n derivatives $f^{(k)}(a)$, $k = 1, 2, \ldots n$, are continuous on an interval containing the points a and x, then $f(x)$ has the following expansion into a polynomial of order n and variable $(x - a)$:

$$f(x) = f_n(x), \tag{A3.7}$$

where

$$f_n(x) = f(a) + \Sigma f^{(k)}(a)(x - a)^k/k! + R_n, \tag{A3.8}$$

where the remainder after n terms is

$$R_n = \int dt (x - t)^n f^{(n+1)}(t)/n!. \tag{A3.9}$$

Here the range of integration is from a to x. In practice, we try to choose n and a so that R_n can be ignored for the range of x of current interest.

Proof of Taylor's theorem

We will prove Taylor's theorem by the method of induction. (A simple example of this method is given in Problem [A3.2].) We suppose that the theorem is true for n, then show that it must necessarily hold for $(n + 1)$.

Since equation (A3.7) claims that $f(x)$ is independent of n, the difference between f_{n+1} and f_n will have to be shown to be zero. If we write out this difference using the preceding postulated equations, we have to find the difference between R_{n+1} and R_n.

► *Exercise* A3.3.1   Integrate R_{n+1} by parts to show that

$$R_{n+1} = R_n + f^{(n+1)}(a)(x - a)^{n+1}/(n + 1)! \qquad (A3.10)$$

☐

Thus, there is no difference between the series (plus remainder) carried out to n terms or to $n + 1$ terms. So, what *is* the value of the series? Consider $n = 0$, since any choice of n will do and this is probably an easy case. We write out equation (A3.8) with its remainder R_0,

$$f_0(x) = f(a) + \int dt(x - t)^0 f^{(1)}(t)/(0)!, \qquad (A3.11)$$

in which the integral is from a to x. On integrating the second term by parts we readily find that it is just $f(x) - f(a)$. Therefore, $f_0(x) = f(a) + [f(x) - f(a)] = f(x)$, just as we had hoped. Hence, although it may not be obvious, Taylor's theorem is proved. Note that this method of proof by induction closely resembles loop structures in programming.

Interpreting Taylor series

Now that the formalism is out of the way, let's see what the Taylor series means.

What is the meaning of a in a Taylor series? Since it is arbitrary, apart from the requirement of continuous derivatives, we can choose it for convenience of analytic and numerical calculation. Analytically, all the derivatives have to be evaluated as functions of x, then evaluated at $x = a$. Thus, the derivatives should be able to be obtained in a straightforward way.

Numerically, we want the power series in $(x - a)$ to converge rapidly because of our appropriate choice of a for the x values of interest. In numerical work we always neglect the remainder term, or at best make a rough estimate of it; after all, if we knew how to work it out exactly, what would be the point of making the messy expansion? We might as well have just worked out $f(x)$ directly!

A Scottish mathematician who became a professor by age 19 (those were bonny days) made the canny choice $a = 0$. We call such a series a Maclaurin series after Colin Maclaurin (1698–1746), who developed the use of Taylor series.

► *Exercise* A3.3.2   To clarify the relation between polynomials and Taylor series, use Taylor's theorem to show that if $f(x)$ is a polynomial of degree p, and if a Maclaurin expansion is made, then the coefficients b_k of the polynomial are given by $b_k = f^{(k)}(0)/k!$, and the remainder $R_p = 0$. ☐

A3.4 {2}   Taylor expansions of useful functions

Many of the equations and approximations in later chapters, especially A4, A6, A7, and A8, require the power series expansions of functions such as the exponential, cosine, and sine. In this section we derive these expansions and discuss some of their properties, with emphasis on their suitability for numerical computation.

Expansion of exponentials

The Maclaurin expansion of the exponential function exp (x) serves as a starting point for many other Maclaurin expansions, so we investigate it first.

► *Exercise*   A3.4.1

(a) Suppose that the Maclaurin expansion of exp (x) were known. What transformation would be most practical to produce rapid convergence of the expansion exp (ax) for x values near some arbitrary a?

(b) Find a fast and accurate way to compute exp (mx) when m is integral and exp (x) is given. This property is important in deriving the fast Fourier transform algorithm in A5.8.

(c) How is the power series expansion of r^x, with r fixed, related to that of exp (x)? □

In deriving the Taylor expansion of the exponential function, the derivatives are especially simple, since for each k,

$$f^{(k)}(x) = \exp(x). \tag{A3.12}$$

The derivatives are to be evaluated with x replaced by a, which is zero. Thus $f^{(k)}(a) = \exp(0) = 1$. Plug this into the Taylor expansion formula (A3.8) to get

$$\exp(x) = 1 + \Sigma\, x^k/k! + R_n, \tag{A3.13}$$

in which the summation over k is from 1 to n. The remainder R_n can be limited by replacing exp (t) in the remainder formula (A3.9) by 3^t, since we know that $e < 3$. The remainder integral can then be done, resulting in $R_n < 3^n x^{n+1}/(n+1)!$. The ratio of successive terms tends to $3x/n$ for very large n, so the series converges. Thus, we can write the Maclaurin expansion of the exponential function as

$$\exp(x) = \Sigma\, a_k x^k, \tag{A3.14}$$

in which the sum now begins at zero. The convergence of the exponential

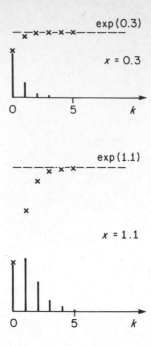

Figure A3.4.1   Convergence of the exponential series for arguments 0.3 and 1.1.

series is illustrated in Figure A3.4.1. The coefficients satisfy the iteration formula $a_k = a_{k-1}/k$ for $k > 0$, with $a_0 = 1$.

The next exercise gets you to test some of the numerical properties of expansion (A3.13) for the exponential function. Problem [A3.5] suggests writing a computer program to evaluate exp (x) by its Maclaurin expansion, with the number of terms used in the series being controlled by user input. The Maclaurin expansion of the logarithmic function, the inverse function to the exponential function, is the topic of Problem [A3.6].

▶ *Exercise* A3.4.2

(*a*) Show by direct numerical evaluation of the exponential series (A3.13) with $x = 2$ that to calculate e to about 4 significant figures requires about the first 10 terms of the Maclaurin series. Ignore the remainder term.

(*b*) Using the same numerical method, but with $x = -2$, find how many terms are needed to get 4 significant figures. Why is there a significant difference between the number of terms in (*a*) and (*b*)?

(*c*) Find a neat way to avoid the extra terms in (*b*) and to avoid subtractive cancellation from the alternating terms in the exponential series for $x < 0$. □

Since the exponential function has so many magical properties (for example, Euler's theorem), it is interesting to obtain its power series expansion by another method.

► *Exercise* A3.4.3

(a) Verify the derivative relation $D_x \exp(x) = \exp(x)$ by term-by-term differentiation of the series expansion (A3.13).

(b) On the other hand, suppose that $\exp(x)$ were defined by the differential equation in (a). Then assume that the exponential function has a power series expansion $\exp(x) = \Sigma\, b_k x^k$; the sum can be assumed to start at $k = 0$. Differentiate the series term-by-term to show that the coefficients must satisfy $b_k = b_{k-1}/k$, $k > 0$. Choose the normalization so that $\exp(0) = 1$, to show that $b_0 = 1$. Thus you have derived the Maclaurin expansion of the exponential function without invoking Taylor's theorem. □

In this exercise we have used the power series method of solving differential equations analytically. We return to the solution of differential equations in A6 and A7.

Series for circular functions

The circular, or trigonometric, functions $\cos(x)$ and $\sin(x)$ occur frequently, especially in the solution of problems describing oscillatory motion. Their Taylor expansions are of great practical importance.

To find the Maclaurin expansion of the cosine function we need the derivatives of the cosine, all evaluated at $x = a = 0$. For x in radians (not degrees) the derivatives are $f^{(k)}(x) = \cos(x)$ if $\mathrm{mod}(k,4) = 0$, $f^{(k)}(x) = -\sin(x)$ if $\mathrm{mod}(k,4) = 1$, $f^{(k)}(x) = -\cos(x)$ if $\mathrm{mod}(k,4) = 2$, $f^{(k)}(x) = \sin x$ if $\mathrm{mod}(k,4) = 3$. Thus, with $a = 0$ the derivatives cycle around the values 1, 0, -1, 0. The terms of the series that have odd powers of x therefore have zero coefficients. The Maclaurin expansion of the cosine function is therefore

$$\cos(x) = \Sigma\, b_k x^k, \tag{A3.15}$$

with $b_k = 0$ if k is odd, and $b_{2k} = -b_{2k-2}/(2k)$, $b_0 = 1$, for the coefficients of even powers of x. Explicitly

$$\cos(x) = 1 - x^2/2 + x^4/24 - x^6/720 + \dots \tag{A3.16}$$

We note that in this formula x must be in radians, not in degrees; $1° = 0.0174533$ radian, or (very roughly) $1° \approx (1/60)$ radian, on using the approximation that $\pi = 3$. The convergence of the cosine series is illustrated in Figure A3.4.2.

We can make immediate practical use of the cosine Maclaurin series:

► *Exercise* A3.4.4   In measuring lengths along nearly parallel lines you will have noticed that a small amount of nonparallelism doesn't give much error in a length measurement. To understand this convenient fact, suppose that an error of 2° is made in estimating parallels. (A good carpenter, surveyor, or drafter can do about

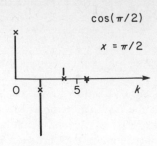

Figure A3.4.2   Convergence of the cosine series for argument π/2, for which the series converges to zero.

this well by eye.) Show that the fractional error in a length measurement is then only about one part in 2000. □

The Maclaurin expansion of the sine function is made similarly to that for the cosine. So I leave it up to you:

▶ *Exercise* A3.4.5   Find (but not in a textbook) the Maclaurin expansion of the sine function. □

If you have done it right, you will have derived the Maclaurin expansion of the sine function

$$\sin (x) \; = \; \Sigma \, b_k x^k, \tag{A3.17}$$

in which the summation over k begins at zero. The coefficients are given by $b_k = 0$ if k is even, and $b_{2k+1} = -b_{2k-1}/(2k + 1)$, with $b_1 = 1$. Explicitly,

$$\sin (x) \; = \; x(1 \, - \, x^2/6 \, + \, x^4/120 \, - \, x^6/5040 \, + \, \dots \,), \tag{A3.18}$$

where, as ever, x is in radians, not in degrees.

Given the power series expansions of the exponential, cosine, and sine functions, one may prove Euler's theorem by comparing these power series. This proof is offered (by the student) as Problem [A3.7].

Numerical methods for computing the cosine and sine functions are suggested in Problem [A3.8].

Hyperbolic function expansions

The hyperbolic functions *cosh* and *sinh* in A2.4 can be expanded in Taylor series. One may use Taylor's theorem directly, or the expansion of the exponential function, or the definition of hyperbolic functions as circular functions of imaginary argument, as in A2.4. Whichever way is used, you should arrive at the same formula for the Maclaurin expansion of the hyperbolic functions:

$$\cosh (x) = 1 + x^2/2 + x^4/24 + x^6/720 + \ldots, \tag{A3.19}$$

and

$$\sinh (x) = x(1 + x^2/6 + x^4/120 + x^6/5040 + \ldots). \tag{A3.20}$$

▶ *Exercise* A3.4.6

(*a*) Derive the series expansions (A3.19) and (A3.20) for the hyperbolic functions.
(*b*) Using the formulas derived in (*a*), explain why the hyperbolic functions (as their name implies) grow indefinitely large as *x* increases, whereas the circular functions are bounded (by the unit circle). ☐

Having considered the very practicable series expansions for the exponential, circular, and hyperbolic functions, we now turn to a very useful approximation formula that is of great value in applicable mathematics.

A3.5 {1}   The binomial approximation

Many problems in applicable mathematics are greatly simplified if they can be approximated by a linear dependence on the variable of interest. The binomial approximation is one way to produce such a linear dependence. Its skillful application can produce an orderly solution out of an otherwise intractable problem.

The binomial approximation derived

The binomial approximation is that

$$(1 + x)^n \approx 1 + nx \tag{A3.21}$$

provided that $|(n - 1)x| \ll 1$. More generally therefore

$$(a + b)^n \approx a^n(a + nb/a) \tag{A3.22}$$

if $|(n - 1)b/a| \ll 1$. Note that in the binomial approximation the power *n* is arbitrary, not necessarily integral.

To derive the binomial approximation, we make a Maclaurin expansion of the function $f(x) = (1 + x)^n$. The required derivatives are very straightforward to obtain, and when inserted into Taylor's formula (A3.8) produce

$$(1 + x)^n = 1 + [1 + (n - 1)x/2]nx + \ldots \tag{A3.23}$$

If n were a positive integer and we carried out this procedure until n derivatives had been taken, we would derive the binomial theorem, an exact expansion. (See Flanders, Korfhage, and Price, Chapter 3.) For any n, inspection of equation (A3.23) shows that the condition given for the validity of approximation (A3.21) is sufficient to give a good approximation, since higher powers of x are thereby also relatively negligible.

Interpreting the binomial approximation

The binomial approximation can be interpreted geometrically when n is a positive integer, as indicated in Figure A3.5.1 for $n = 2$ (squares) and $n = 3$ (cubes).

► *Exercise*   A3.5.1   According to the binomial approximation (A3.21) the approximation holds for any value of n. Can a drawing be made that illustrates it for (*a*) $n = 4$, and (*b*) $n = 2.5$? ☐

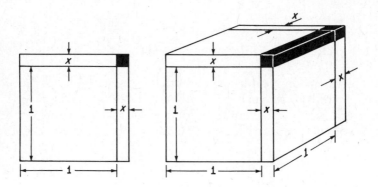

Figure A3.5.1   Binomial approximation illustrated for squares and cubes. Ignored areas and volumes are black. For the cube, two of the ignored volumes are hidden at the back.

Applying the binomial approximation

To illustrate the power of the binomial approximation, consider the following numerical examples:

(1) $(1.01)^{10} \approx 1 + 10*0.01 = 1.1$, whereas, to 6 significant figures $(1.01)^{10} = 1.10462$, so that the binomial approximation is in error by less than 5 parts in 10^4.

(2) $(2.01)^{-3} \approx 2^{-3}[1 + (-3)0.01/2] = 0.985/8 = 0.123$, whereas, $(2.01)^{-3}$ to 6 significant figures is 0.123144.

Now here are some more challenging examples for you to try:

▶ *Exercise* A3.5.2 Estimate by the binomial approximation and compare with exact values; $(2.01)^{10}$, $(2.1)^{21}$, the ratio of $(2001)^n$ to 2000^n. □

These examples show how much more difficult, for little gain in accuracy, the exact calculations are.

The binomial approximation is most useful when a linearizing approximation is needed, as in some of the numerical approximation methods in A4, and in estimating error terms in the solution of differential equations in A6 and A7. Here is an interesting application from the theory of heat transfer.

▶ *Exercise* A3.5.3 Heat transfer by radiation is a very important consideration in energy conservation practice and in energy dissipation in situations where radiation is the only heat exchange mechanism. A topical example is cooling the external components of a space vehicle exposed to direct solar radiation.

Stefan's law of black-body radiation states that the energy radiated at absolute temperature T is proportional to T^4.

(*a*) Use the binomial approximation to show that the net radiant energy transfer between two bodies, one at T, the other at $T + \Delta T$, where ΔT is small compared to T, is proportional to $T^3 \Delta T$.

(*b*) Two bodies (animate or inanimate) have a temperature difference of 10°C. By what factor is the radiative transfer between them increased when the temperature of the cooler one is increased from 27°C to 127°C? □

After these examples of analytical and numerical methods for power series, you would probably be glad of a diversion. The following section should be interesting.

A3.6 {1} Diversion: Financial interest schemes

Banks, credit-card companies, and loan sharks depend (to varying degrees) on the confusion between different schemes for charging interest on financial loans. The mathematics divulged in the preceding sections are sufficient for us to appreciate and estimate the differences between simple interest, quarterly-compounded interest, and daily-compounded interest.

We consider a repayment time of t years at an annual interest rate quoted as r, and applied to a principal of unit funds. By r is meant the fraction I_s of the principal that would be owed after 1 year at simple interest;

$$I_s = (1 + rt) - 1 = rt. \tag{A3.24}$$

Note that interest rates are usually quoted as percentages, so that a 15% interest rate has $r = 0.15$. The general formula to calculate compound interest if it is paid n times annually is thus $I = (1 + rt/n)^n - 1$. According to the binomial approximation (A3.21), this reduces to simple interest, $I \approx rt$, if $(n - 1)rt/n \ll 1$, that is, $rt \ll 1$.

If interest is accrued quarterly, then $n = 4$, so that the interest repayment after t years, or fraction thereof, is

$$I_Q = (1 + rt/4)^4 - 1 = rt[1 + 3rt/8 + (rt)^2/16 + (rt)^3/256]. \quad (A3.25)$$

Quarterly interest thus exceeds simple interest by the fraction $(I_Q - I_s)/I_s = 3rt[1 + rt/6 + (rt)^2/96]/8$. Note that in the binomial approximation only the first term in the expansion of I_Q is retained.

If interest charges accrue daily, then the interest due is I_D, where

$$I_D = (1 + rt/365)^{365} - 1, \quad (A3.26)$$

(except on leap years) and only the first term beyond the binomial approximation is enough;

$$I_D \approx [1 + 365rt/365 + 365*364(rt/365)^2/2] - 1 \leqslant rt + (rt)^2/2. \quad (A3.27)$$

Thus, daily interest exceeds simple interest by a fraction of about $rt/2$, which is nearly $rt/8$ greater than quarterly interest.

As $n \rightarrow \infty$ the series for $I + 1$ approaches the Taylor series for the exponential, that is,

$$(1 + rt/n)^n \rightarrow \exp(rt), \quad (A3.28)$$

as is hinted at in the formula for I_D. Thus, interest accruing at each instant is due at an amount

$$I_E = \exp(rt) - 1 = rt + (rt)^2/2 + (rt)^3/6 + \dots \quad (A3.29)$$

This is only negligibly greater than daily interest; but it is certainly not good business for a lender to advertise "exponentially increasing" interest charges.

Readers majoring in science might consider this financial diversion below their lofty concern, and more suitable to penny-pinching students in business administration. But did you know that Isaac Newton, who was Keeper of the Mint, was the first scientist (using differential calculus) to pose and solve this problem of exponential interest?

Problem [A3.9] illustrates by a topical example the differences between various interest schemes.

Problems with power series

[A3.1] {1} (Complex geometric series) In the geometric series, suppose that r is a complex number, $r = |r| \exp(i\theta)$ with θ real. Generalize the results in A3.2 to show that the geometric series converges if $|r| < 1$ and diverges if $|r| \geqslant 1$.

[A3.2] {2}   (Proof by induction) This problem illustrates use of the method of proof by induction to verify the formula for the partial sum of a geometric series. Suppose that we suspect the formula for the partial sum to n terms to be $S_n = a(1 - r^n)/(1 - r)$. We might have arrived at this result by explicitly summing the series for low values of n. Use the method of induction to prove this result, by first showing that its truth for n implies its truth for $n + 1$, that is $S_n + ar^n = S_{n+1}$. Then note that $S_1 = a$ to show that the result holds for any positive integer n.

This method of induction is logically analogous to iterative methods for generating successive terms in a power series and to loop structures in computer programs.

[A3.3] {1}   (Numerical geometric series)
(*a*) Write and test a program for computer or pocket calculator to calculate and sum directly the first n terms of a geometric series. Use the iteration formula in a loop structure to generate successive terms in terms of previous terms and to sum the partial series.
(*b*) Add a section to the program to check numerically how closely the results from (*a*) agree with the analytic formula for S_n. Do this for values of r that might prove troublesome; which values are these? (If you don't know the answer at first, use the program for a variety of values of r to help indicate the appropriate values.)

[A3.4] {2}   (Series by analogy) We discover another way to sum the geometric series by using an analogy with the problem of a bouncing ball. Suppose that a ball has already bounced n times, then it bounces m times more. Show, without using the closed expression for S_n (which is what you are trying to find), that the relation between the total vertical distances traveled must be $S_{n+m} = S_n + r^n S_m$, with $S_1 = a$. Now set $n = 1, m = N - 1$ to get a relation between S_N and S_{N-1}. Then set $n = N - 1$ and $m = 1$ to get another relation. From these two relations show directly that $S_N = a(1 - r^N)/(1 - r)$.

[A3.5] {2}   (Numerical exponentials)
(*a*) Write a program for the numerical evaluation of the exponential function from its Maclaurin expansion. The program should read the argument x and the value ε, the fractional accuracy that the power series result should have. Use the iteration formula given below (A3.14) for the coefficients of successive terms in the series for exp (x).

The program should handle both negative and positive values of x. After output of the result, the program should loop to obtain new input values.
(*b*) Add a section to your program to compare your series expansion with that obtained from the program library function.
(*c*) Find out what algorithm is used to evaluate the exponential function in

your computer or calculator; why is the power series expansion not used? (Hopefully, it is not.)

[A3.6] {3} (Logarithms in series)
(a) Expand ln $(1 + x)$ in a Maclaurin series to show that, for $|x| < 1$, ln $(1 + x) = \Sigma (-1)^{k+1} x^k/k$, in which the sum starts at $k = 1$.
(b) As an alternative method, use the geometric series result (A3.6) with $r = -x$ and integrate both sides of this expression to obtain the formula in (a). Note that this method assumes (correctly) that summation and integration can be interchanged for convergent power series. (See Rainville Chapter 8.)
(c) Consider a table of logarithms evaluated at points $t - h, t, t + h, ...,$ and the problem of interpolating to some point a distance d from t. Clearly, $|d| < h/2$ is sufficient. Show that the error in the approximation ln $(t + d) = $ ln $(t) + [1 - d/(2t)]d/t$ does not exceed $(h/t)^3/24$. For example, if $t = 1$ and $h = 0.01$, then the error is less than 10^{-7}.

Thus, if logs are required over a very small range, a table of about 100 entries will suffice, with the interpolation formula just presented, to give values that will typically be as accurate as direct evaluation of the logarithm by some other algorithm. The interpolation method will almost certainly be much faster.

[A3.7] {1} (Euler and Maclaurin) Prove Euler's theorem (A2.38) by considering the Maclaurin series for exp (ix), cos (x), sin (x). Sum the series for cos $(x) + i$ sin (x) term by term (which is legitimate for convergent series). Show that you get the power series for exp (ix). This provides an alternative proof to that given in A2.3.

[A3.8] {2} (Numerical cosines and sines) Consider the numerical evaluation of the cosine and sine functions. Since cos $(\pi/2 - x) = $ sin (x), cos $(-x) = $ cos (x), sin $(-x) = -$ sin (x), either power series need be used only for $0 \leqslant x < \pi/4$. Indeed, one can do better than this by using the trigonometric formulas for half angles, so that only $|x| < \pi/8$ need be evaluated.
(a) Show that the first 6 terms of the Maclaurin series are sufficient (if round-off error is negligible) to get the cosine and sine to about 7 significant figures.
(b) Show that the cosine series to 6 terms can be written compactly as

$$\cos (x) = 1 - (1 - (1/6 - (1/90 - (1/2520 - y/113400)y)y)y)y, \quad (A3.30)$$

where $y = x^2/2$. This factorized algorithm for evaluating a polynomial is called *Horner's method*. Why is this form convenient and fast for computer coding?
(c) After all this theorizing, write a program to check out these results for the

cosine function. Compare your power series expansion using Horner's method against whatever library function your calculating engine provides. If your computer system has a convenient and accurate timing method, compare this power series formula for accuracy and speed with the computer system's method.

For a clear discussion of efficient methods of computing common functions see Cody and Waite.

[A3.9] {1}   (Financing solar energy) A budding inventor of solar energy technology products mortgages her house for a government development loan of 30,000 dollars at 8%.

(*a*) How much interest would she owe after 1 year at simple interest?

(*b*) How much more would she owe after 1 year of quarterly compound interest?

(*c*) Suppose that the interest charges accrue daily. About how much more does she owe after 1 year than if she were charged simple interest at 8%?

(*d*) She needs another 10,000 dollars to market her product to an energy-starved but skeptical business world. The Flim-Flam Loan Company tries to acquire her loan by claiming that exponential interest is more advantageous for her than daily interest. They offer her 16% (quoted as if it were simple interest). How much more would she owe them after 1 year?

References on power series

Cody, W. J., and W. Waite, *Software Manual for the Elementary Functions*, Prentice-Hall, Englewood Cliffs, N.J., 1980.

Flanders, H., R. R. Korfhage, and J. J. Price, *A Second Course in Calculus*, Academic Press, New York, 1974.

Hansen, E. R., *A Table of Series and Products*, Prentice-Hall, Englewood Cliffs, N.J., 1975.

Jolley, L. B. W., *Summation of Series*, Dover, New York, 1961.

Maurone, P. A., "Time of Flight for the Bouncing Ball Experiment," *American Journal of Physics*, **47**, 560 (1979).

Rainville, E. D., *Infinite Series*, Macmillan, New York, 1967.

A4 NUMERICAL DERIVATIVES, INTEGRALS, AND CURVE FITTING

From this chapter on numerical derivatives, integrals, and curve fitting, you will learn techniques appropriate to digital computer analysis. Most of the techniques are straightforward to apply and are of general applicability. An important topic, not usually presented at this level, is spline fitting of curves. This chapter develops the numerical analysis formulas and algorithms, with the computing laboratories extending them to practical applications.

Looking ahead in this chapter, we see that section A4.1 emphasizes the distinction of numerical mathematics from pure mathematics, especially in the discreteness of data. In A4.2 the topic is numerical noise as it appears in round-off and truncation errors in arithmetic, subtractive cancellation, and in unstable problems and methods. Formulas for the approximation of derivatives are derived in A4.3, although most of their applications appear in computing laboratory L3. The important topic of numerical integration is introduced in A4.4, where we confine our attention to the trapezoid and Simpson formulas. These formulas are applied and extensively tested in L5 on calculating electrostatic potentials by integration. As a diversion, we briefly review analytic evaluation by computer in A4.5.

Spline fitting of data is the topic of A4.6, where a basic algorithm is derived. The coding of this algorithm and its applications are given in L7. The least-squares principle is introduced in section A4.7. It serves as the basis for many types of expansions, including the Fourier expansions in A5. Least-squares fitting formulas are derived in A4.7, with extensions in computing laboratory L8, where straight-line fitting with errors in both variables is considered and programs with exercises are presented.

In addition to many exercises in the text, this chapter has extensive problems and further readings in numerical analysis, usually going beyond the topics considered here, given at the end.

A4.1 {1}   The discreteness of data

It is a fact that data are always obtained at discrete values of the independent variables. Although we may believe that the quantity being measured is a smooth function of the parameters (independent variables), we believe rather than know. For a good example of how willing disbelief can lead to significant results, read about the discovery of the quark-antiquark resonance states, which led to a Nobel prize in physics. This is described in ''Discovery of Massive Neutral Vector Mesons'' in the book edited by Maglich. This example of resonance in nature required that the experimenters obtain data at very small increments of energy, steps much smaller than reasonable by then current understanding of quark physics.

Numerical mathematics

In numerical evaluation of formulas, discrete values of the variables are also required, rather than the continous range of values often allowed by the defining formulas. For practical reasons, such as minimizing computer storage and time, one often characterizes a function by calculating accurately as few values as possible with large steps in the independent variables, then finds related properties of the function such as its values at intermediate values of the variables (interpolated values), its slope (derivative), and the area under the function between two limits (integral) by approximate numerical procedures.

The scientist is often concerned with determining rates (derivatives) and totals (integrals) from data. Therefore, experimental research often requires approximation of derivatives and integrals from data at discrete points. In theoretical work the analytic formula for an integral may not be known, but we can nearly always find a good numerical approximation to it by arithmetic computation.

Discreteness

The emphasis in this chapter and much of this book is quite different from those in pure mathematics, in which existence of quantities is of major concern and analytic solutions to derivatives and integrals are usually emphasized. However, we must always be guided by such mathematics, especially on questions of existence and convergence of quantities that we attempt to evaluate numerically.

The present branch of mathematics is called *numerical analysis*, and the process of replacing continuous functions by discrete values is called *discretization*. A table of numerical values of a hyperbolic function, as discussed in A2.4, is an example of a discretized version of these functions over a limited range and increment of the argument.

A4.2 {1}   Numerical noise

In a digital computer, most numbers can be represented to only a finite number of significant figures. For example, in the usual binary computer using base-2 arithmetic, if there are b bits in each computer word (single variable), then only numbers that can be factored into no more than b adjacent powers of 2 can be exactly represented. For example, in a binary computer the number 0.1 can not be exactly represented.

One speaks of "computer arithmetic," in which the number of digits is finite, in distinction to formal or exact arithmetic. The errors in calculations introduced by the finite number of significant figures carried by the computer must be clearly distinguished from the *random* error rate in the repeatability of the identical arithmetic operation. Modern digital computers are designed so that the random error rate in arithmetic operations is typically 1 in 10^{12}, which is about one arithmetic error in 10 days. It is perhaps surprising, except to evolutionists, that the errors of replication in natural organisms such as the bacterium *Escherichia coli* are much greater than this, as you can read about in the article by Gueron.

The two main methods of handling the least-significant digits after an arithmetic operation are *round-off* and *truncation*. We now define and give examples of these sources of numerical noise.

Round-off error

When round-off is used, the least-significant digit is rounded to the nearest least-but-one significant digit. For example, 0.12345676 rounded to 7 significant figures is 0.1234568, while 0.12345674 rounded to 7 significant figures is 0.1234567; the error due to such round-off is therefore about 10^{-7} The numbers 1.99999 and 2.00001 are both 2.0000 when rounded to 5 significant figures.

▶ *Exercise* A4.2.1   A certain personal computer has 16 bits (binary digits) to represent each floating-point real number, including its sign. Estimate the fractional accuracy that a rounded number has in this computer. Give the answer in decimal notation. ☐

In large general-purpose computers (macrocomputers) about 7 significant decimal digits are retained after each calculation involving $+$, $-$, $*$, or $/$. Results are therefore limited to less than 7-figure accuracy. Such computers may also be capable of double-precision arithmetic, in which at least twice this number of significant figures is retained. Computers built primarily for scientific applications, such as supercomputers, usually have round-off errors in the 16th significant decimal digit. Thus, it is risky to transfer a numerical program directly from a double-precision system to a single-precision system without making careful checks of the changes in accuracy of results that may entail.

As an example of the effects of round-off noise, consider the iteration formula $x_0 = 0$, $x_k = x_{k-1} + h$ for $k > 0$, which is to be used as an alternative to the algebraically equivalent formula $x_k = kh$, which we call $x_k(mult)$. In some computers, addition is faster than multiplication, so that time would be saved by using the iteration formula to produce a value $x_k(add)$. However, after about k steps the round-off error will be about $\sqrt{k}$ times that after each step, if we assume that the round-off process can be described as a random walk, as discussed in L6. For example, $x_{100}(add)$ has about one fewer significant decimal digits than $x_{100}(mult)$. If the computer times for addition and multiplication are similar, then the multiplication formula is probably superior.

► *Exercise* A4.2.2   (Testing round-off error)

(a) Use a small computer or calculator program with say $h = 0.1$ and output the error estimator $e_k = [x_k(mult) - x_k(add)]/x_k(mult)$ for a range of k values. Does e_k increase roughly as $\sqrt{k}$.?

(b) Why is it inaccurate to use the form $e_k = 1 - x_k(add)/x_k(mult)$ to determine e_k? □

In all the exercises in *Computing in Applied Science* 7-figure accuracy is sufficient. If this does not appear to be so, then your algorithm, or your coding (or both), are probably not well designed.

Truncation error

In truncation, sometimes called "chopping," the least-significant digit is dropped. For example, 0.1234567 truncated to 6 significant figures is 0.123456, 3.9999 truncated to 1 significant figure is 3, while 4.0001 truncated to 1 significant figure is 4. The last two examples occur when a computer program determines the integer part of a number. Some instructional programming languages automatically add a number of order of the expected round-off error before truncating to an integer. This number is called a "fuzz." However, this is risky, and it is probably better to code an appropriate fuzz oneself. (It's too easy to get nabbed by the fuzz.)

► *Exercise* A4.2.3   (Computer arithmetic breaks the rules)

(a) Demonstrate by computer or calculator examples that finite-precision arithmetic does not satisfy the associative law of addition $(a + b) + c = (b + c) + a$. Cases in which the signs of a, b, c differ should be exhibited.

(b) Prove that for addition the optimal ordering of numbers to produce the smallest error from loss of significant figures is to add the numbers arranged by increasing order of their magnitude.

(c) Demonstrate that the identity $(a/b)b = a$ for $b \neq 0$, is not always satisfied in computer arithmetic. For what ranges of a, b, c are the relative errors largest in your computing system? □

A consequence of the nonassociative nature of addition in computer arithmetic is that the result $s = a + b + c$ depends on the order of evaluation, left-to-right as $(a + b) + c$ or right-to-left as $(c + b) + a$. This is demonstrated in Exercise A4.2.3. To determine the order of evaluation for a pocket calculator you may have to use trial and error on sums sensitive to round-off or truncation. For a computer program you will need to know how the compiler program translates your program into direct computer instructions. Problem [A4.1] investigates further optimal schemes for minimizing the effects from loss of significant figures.

Unstable problems and unstable methods

We have just considered the effects of finite accuracy of computer arithmetic. These must be distinguished from unstable (ill-conditioned) problems. An *unstable problem* is one in which even the exact solution is very sensitive to slight changes in the data. An example of an unstable problem is the solution of the pair of linear equations

$$x + y = 2 \tag{A4.1}$$

$$0.99x + 1.01y = 2, \tag{A4.2}$$

whose algebraic solution is $x = 1$, $y = 1$. However, the solution to a very similar pair of equations

$$x + y = 2 \tag{A4.3}$$

$$0.99x + y = 2.01 \tag{A4.4}$$

is exactly $x = -1$, $y = 3$. If such an unstable problem is attempted to be solved numerically, then the results will be very sensitive to numerical noise.

The unstable problem of solving two linear equations can be illustrated graphically by plotting the lines corresponding to equations (A4.1) through (A4.4), as shown in Figure A4.2.1. This geometric interpretation of unstable linear equations is developed in Gear, Chapter A10.

For those familiar with determinants, we note that the origin of this sensitivity is that the determinant of the coefficients on the left-hand side is very much less than the values appearing on the right-hand sides of equations (A4.3) and (A4.4). The problem is also related to that of stable or unstable equilibrium in mechanical systems and to noise amplification in electronics. Unstable problems also occur in solving differential equations numerically, especially for stiff differential equations, discussed in A7.7.

We distinguish unstable problems from *unstable methods*. For the latter, the method used to produce the arithmetic solution is unstable, for example, because of sensitivity to round-off errors, even though the exact solution to the problem is insensitive to changes in the data, that is, even though the problem is stable. Unstable methods are especially troublesome when solving differential equations.

Subtractive cancellation

A major source of numerical errors is *subtractive cancellation*. The preceding discussion and exercise will have made it clear that if two numbers are nearly

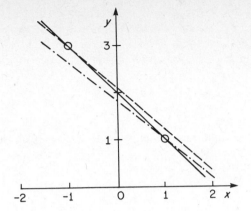

Figure A4.2.1 Lines corresponding to (A4.1) or (A4.3) shown solid, to (A4.2) shown dash-dot, and to (A4.4) shown dashed. The intercepts (within circles) are very sensitive to differences in slope.

equal, then when they are subtracted the result will have a relatively large error.

Consider, for example, a set of numbers $(x_1, x_2, \ldots, x_n)$ whose mean value (average) x_{av} is given by

$$x_{av} = (\Sigma\, x_j)/n, \tag{A4.5}$$

in which the sum over j is from 1 to n. The variance of this set is V, given by

$$V = \Sigma\, (x_j - x_{uv})^2, \tag{A4.6}$$

in which the sum is over the n data points. One method of calculating V consists of expanding the squares in the last equation, interchanging the summation order, then using the definition of the average to obtain

$$V = \Sigma\, x_j^2 - x_{av}^2, \tag{A4.7}$$

again summing over all data points. This expression is algebraically equivalent to the defining equation for V, (A4.6). Expression (A4.7) is often used, especially in pocket calculators, to speed up calculations of variance because the sums over x_j and x_j^2 and the values of x_j need not be stored in memory. However, in the second form for V, subtractive cancellation inevitably occurs because the terms are all positive. For example, if all the values in the set are equal (and therefore equal to their mean) and large, V will be zero but the intermediate sums will be very large. To further complicate the numerical problems of the second form, if the x_j are of order of or larger than the square root of the largest number that can be represented in the computer, then over-flow errors will occur in evaluating the sum of squares. Such errors are much less likely when the first form for V is used.

As another example of subtractive cancellation, consider the equation

$$(x + 1)(x + 10^{-8}) = 0. \tag{A4.8}$$

This has exact solutions $x = -1$, $x = -10^{-8}$, and the problem of finding the solutions is a stable problem. However, the equation could be expanded into the standard quadratic form

$$ax^2 + bx + c = 0, \tag{A4.9}$$

which has algebraic solutions

$$x = [-b \pm \sqrt{(b^2 - 4ac)}]/2a. \tag{A4.10}$$

Here $a = 1$, $b = 1.00000001$, $c = 10^{-8}$. When I substituted these values into equation (A4.10) and used my pocket calculator with 8 significant figures, it produced the root $x = -1.000000$, and the root $x = 0$, which is inaccurate in the first significant figure. This subtractive cancellation occurred because b and the square-root term are nearly equal. Thus, the algorithm given by the last equation is an unstable method if $b^2 \gg 4ac$. A remedy for this numerical instability due to subtractive cancellation is suggested in Problem [A4.3]. The problem appears also in solving for the slope of a least-squares-fit straight line with errors in both variables, as in L8.1.

The term *truncation* is also used in reference to power series. When the remainder after n terms of a Taylor expansion (considered in A3) is ignored, then one has an nth order polynomial approximation.

Further explorations of the problem of numerical noise are suggested in Problem [A4.2]. In computing laboratory section L6.2 there are several exercises in which you may use Monte-Carlo simulation to investigate round-off errors.

Numerical evaluation of polynomials

Polynomials occur so frequently in applicable mathematics that it is worth-while to consider the best methods for evaluating them. Suppose that the polynomial is

$$p_n(x) = \Sigma a_k x^k, \tag{A4.11}$$

where the sum over k runs from zero to n. Assume that the $n + 1$ coefficients a_k are given numerically, as in a table. There are two considerations affecting the choice of a method; accuracy then speed. The direct evaluation of (A4.11) requires n additions. If the powers of x are generated by iteration, $x_k = x x_{k-1}$, then $2n - 1$ multiplications are needed to evaluate $p_n(x)$. A suitable algorithm is

```
p : = a₀; xₖ : = 1; Initialize p and x powers
    Do the next two steps n times
    xₖ : = x(xₖ); Powers of x by iteration
    p : = p + aₖ*xₖ; Add in (k)th term to polynomial
```

Alternatively, the number of multiplications can be nearly halved by using Horner's method, for which the algorithm, written in pseudocode, is

```
p : = aₙ
    Do the next step n times
    p : = aₖ + p*x
```

In Horner's method the round-off error is decreased and the speed is increased. As an example of Horner's algorithm we have: $p_3(x) = a_0 + (a_1 + (a_2 + a_3 x)x)x$. Problem [A4.4] compares the two methods for polynomial expansions of $\tan^{-1}(x)$. Extensive discussions, examples, and exploratory Fortran programs on numerical noise in computer calculations are given in the first chapter of the numerical analysis text by Maron.

A4.3 {1}   Approximation of derivatives

In numerical analysis, differentiation of a function is usually much more difficult than integration, because of the great sensitivity of differentiation to subtractive cancellation. This fact is in contrast to the situation in formal analysis (commonly just called "analysis"), in which most functions can be differentiated but relatively few can be integrated.

For numerical derivatives it is important to find stable methods by developing algorithms that are insensitive to round-off errors. We investigate this in the following.

Forward-difference derivatives

One way to define the first derivative of a function $y(x)$ at the point x is

$$y'(x) = \lim D_+(h), \qquad (A4.12)$$

where the limit is taken as $h \to 0$, of the quantity

$$D_+(h) = [y(x + h) - y(x)]/h. \qquad (A4.13)$$

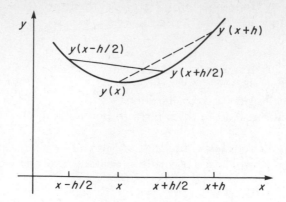

Figure A4.3.1   Numerical first derivatives by forward differences (dashed line) and by central differences (solid line). The step size has been chosen to emphasize the different slopes.

This formula gives the finite-difference approximation to $y'(x)$. A geometric interpretation of this formula is shown in Figure 4.3.1. This suggests that with accurate computation of the function y and a small enough value of the step size h, a sufficiently accurate approximation to the derivative may be obtained. Try the following exercise to test this simple idea:

▶ *Exercise* A4.3.1   Write and run a simple program that calculates the finite-difference approximation to the first derivative of $y(x) = \sin(x)$ for a range of x values. For each x value the program should choose 8 successively smaller h values, decreasing by a factor of 10 each time, starting at $h = 0.1$. The program should compare these finite-difference approximations to see how they differ from the exact first derivative $\cos(x)$. □

In this exercise (which I'm sure you completed) the agreement with the exact derivative at first improves as h decreases, then (depending on the number of significant figures in your computer) becomes drastically worse, in complete disagreement with the result from analytical mathematics. With a little thought, or insight into this result, you will realize that the problem is due to subtractive cancellation in the numerator of (A4.13).

In this example we used the forward-difference derivative $D_+(h)$, in which the argument for evaluating the function is always greater than (forward of) or equal to x. Why not try a more balanced approach?

Central-difference derivatives

Consider the equally valid formula for the first derivative

$$y'(x) = \lim D_0(h),\tag{A4.14}$$

where the central-difference first derivative is defined by

$$D_0(h) = [y(x + h/2) - y(x - h/2)]/h. \tag{A4.15}$$

The geometric interpretation of this formula is shown in Figure A4.3.1.

▶ *Exercise* A4.3.2   Write a program to calculate central-difference first derivatives. Compare these values over the same ranges of x and h with the forward differences and with the analytical derivative. A convenient test function is $y(x) = \sin(x)$. Computing laboratory L3 provides suggestions for the detail of such a program. ☐

From this exercise you will find that much larger values of h give an accurate approximation to the derivative. The values of h will probably be large enough that no significant subtractive cancellation errors are likely.

Error estimates for numerical derivatives

The foregoing results can be investigated analytically by using Taylor series, as studied in A3. We are also led to estimates of the errors in each approximation of the first derivative. In the basic formula (A3.8) $x + h$ is used as the variable, and the point at which the derivatives are evaluated is x (replacing a.) With these changes, we have

$$y(x + h) = y(x) + y'(x)h + y''(x)h^2/2 + y'''h^3/6 + y^{iv}h^4/24 + \dots \tag{A4.16}$$

The derivatives are all to be evaluated at x.

The forward-difference derivative formula (A4.13) yields directly

$$D_+(h) = y'(x) + y''h/2 + y'''h^2/6 + \dots \tag{A4.17}$$

Thus, the error in this derivative estimate is of order $y''(x)h/2$.

The central-difference derivative formula (A4.15) requires replacing h by $h/2$ in the Taylor expansion formula. One obtains

$$y(x \pm h/2) = y(x) \pm y'(x)h/2 + y''(x)h^2/8 \pm y'''h^3/48 + \dots \tag{A4.18}$$

Thence, with a little algebra, the central-difference first derivative quantity is

$$D_0(h) = y'(x) + y'''(x)h^2/24 + \dots \tag{A4.19}$$

This has an error of order $y'''(x)h^2/24$. Is this usually significantly smaller than the error in the forward-difference formula?

For example, let us compare the forward and central formulas for the function $y(x) = \sin(x)$ at $x = \pi/4$, with $h = 0.1$. We readily find that $y'(\pi/4) = 1/\sqrt{2}$, $y''(\pi/4) = -1/\sqrt{2}$, $y'''(\pi/4) = -1/\sqrt{2}$. The error estimate for the central-difference derivative is about $-1.5*10^{-4}$, about 200 times smaller than that in the forward-difference derivative.

Computing laboratory L3 has extensive investigations and comparisons of numerical methods for first and second derivatives. Also considered there are methods for extrapolating to the limit $h \to 0$ in numerical approximation of derivatives.

Numerical second derivatives

The numerical estimation of second derivatives clearly requires much more care than for first derivatives. The use of central differences is indicated, and at least three values of y are needed. We use

$$y''(x) = \lim D_0^2(h), \qquad (A4.20)$$

where

$$D_0^2(h) = [y(x + h) - y(x) - (y(x) - y(x - h))]/h^2. \qquad (A4.21)$$

▶ *Exercise* A4.3.3   Use the Taylor expansion about the point x of each function on the right-hand side of (A4.21) to show that

$$D_0^2(h) = y''(x) + y^{iv}(x)h^2/12 + \dots \qquad (A4.22)$$

☐

This result is just that obtained by taking the central-difference first derivative of the two central-difference first derivatives evaluated at $x \pm h/2$.

▶ *Exercise* A4.3.4   Show that the only linear combination of $y(x + h)$, $y(x)$, $y(x - h)$ that has $y(x)$ and $y'(x)$ missing, and that has $y''(x)$ with unit coefficient, is that given in equation (A4.21) for the central-difference second derivative. ☐

Note that subtractive cancellation errors will be increased if (A4.21) is algebraically simplified to $[y(x + h) + y(x - h) - 2y(x)]/h^2$. Unfortunately, most numerical analysis guides write the formula in this way, even though the number of arithmetic operations is not decreased thereby.

▶ *Exercise* A4.3.5
(*a*) The function $y(x) = \sin(x)$ satisfies the differential equation $y''(x) + y(x) = 0$, and the inequality $|\sin(x)| \leqslant 1$. Test the accuracy of the second-derivative formula by checking the differential equation for a range of x values and a range of h values, say $h = 0.1$ and $h = 0.01$.
(*b*) For $y(x) = \exp(x)$ the differential equation is $y''(x) - y(x) = 0$. Investigate numerically the accuracy of the central-difference formula (A4.21). ☐

In computing laboratory L3 there are extensive examples and exercises on numerical estimation of first and second derivatives. The spline-fitting method, considered in A4.6 and L7, also provides a way to compute these derivatives. The topic of numerical solutions of differential equations is developed in A6.6, A7.6, A7.7, and in laboratory L11.

A4.4 {1}   Numerical integration

In this section we introduce numerical integration techniques by deriving formulas for the trapezoid rule and Simpson's rule. We also use the Taylor expansions from A3.3 to investigate the errors in each formula. The two numerical integration formulas are applied extensively in computing laboratory L5 (electrostatic potentials by integration) and in L10, where they can be used in computing the Fourier integral transforms from the ammonia inversion resonance data. An alternative integration method, by Monte-Carlo simulation, is shown in L6.2. The process of integration is also termed *quadrature*, because one is finding the quadrilateral that has the same area.

Integration approximations using constant step size h are readily derived by using a Taylor expansion of the integrand within the region to be integrated. Suppose that the integration is centered on x_0 so that the appropriate Taylor expansion is, using the results of A3.3,

$$y(x) = y(x_0) + y'(x_0)(x - x_0) + y''(x_0)(x - x_0)^2/2 + \dots \quad (A4.23)$$

Note that all the derivatives are evaluated at x_0. The indefinite integral over x is obtained by term-by-term integration of the series (A4.23). The result is readily seen to be

$$\int dx\, y(x) = y(x_0)x + y'(x_0)(x - x_0)^2/2 + y''(x_0)(x - x_0)^3/6 + \dots \quad (A4.24)$$

Each derivative included in this equation requires that y be known at one more x value. With a minimum of two values, (A4.18) can be used to solve for $y(x_0)$ and to substitute into equation (A4.24). This results in the formula for the integral from $x_0 - h/2$ to $x_0 + h/2$

$$\int dx\, y(x) = h[y(x_0 - h/2) + y(x_0 + h/2)]/2 - h^3 y''(x_0)/12 + \dots \quad (A4.25)$$

▶ *Exercise* A4.4.1   Complete the steps leading to this integral result. □

Trapezoid formula

If the second-derivative and higher-order terms in equation (A4.25) are neglected, we obtain a convenient simple formula, the trapezoid formula. We rewrite it, setting $a = x_0 - h/2$, to produce the standard form for the trapezoid formula for the integral from $x = a$ to $x = a + h$

$$\int dx\, y(x) \approx h[y(a) + y(a + h)]/2, \qquad (A4.26)$$

with an error estimate $-h^3 y''(a + h/2)/12$. The term "trapezoid" is used because the integral approximation corresponds to treating $y(x)$ as linear in the interval a to $a + h$. This is indicated in Figure A4.4.1 by the dashed lines.

For a sequence of n equally spaced intervals, additivity of the integration process leads directly to the composite trapezoid rule for the integral from $x = a$ to $x = a + nh$

$$\int dx\, y(x) \approx h[-(y(a) + y(a + nh))/2 + \Sigma y(a + jh)]. \qquad (A4.27)$$

Here the summation over j ranges from $j = 0$ to $j = n$. Note that there are $n + 1$ values of y needed to cover the n intervals.

In the trapezoid formula (A4.27) the function $y(x)$ is approximated as linear in each subinterval of length h, but it may have different slopes in different subintervals, as seen in Figure A4.4.1. The error estimate for the composite trapezoid formula can be derived as (Vandergraft, section 5.2.3) $-nhy''/12$, in which the second derivative is an upper limit to that in the range of integration a to $a + nh$.

The trapezoid formulas will be accurate only in so far as second and higher derivatives are relatively negligible compared with the integral in the region of integration. However, the error in the trapezoid formula decreases by nearly an order of magnitude (1/8) for each halving of h, that is, for each increase in computation time by only a factor of 2. This is a net gain in accuracy/time of a factor of 4.

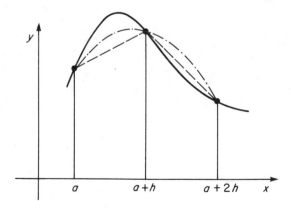

Figure A4.4.1   Trapezoid approximation (dashed lines) and Simpson approximation (dash-dot curve) used in integrating a function (solid curve).

We note that if the trapezoid rule were used in reverse, then we might sum a series approximately in terms of an integral. This is the topic of Problem [A4.6].

The Simpson formula

Integration formulas of increasing accuracy can be developed by including higher-order derivative terms in the basic integration formula (A4.24). For example, by including second derivatives from (A4.22) we find that the integral from $x = x_0 - h/2$ to $x = x_0 + h/2$ is given by

$$\int dx\, y(x) \approx h[y(x_0 - h/2) + 4y(x_0) + y(x_0 + h/2]/6 - (h/2)^5 y^{iv}/90 + \ldots \quad (A4.28)$$

This is a very practical formula because we note that an accuracy/time gain of 16 is obtained for each halving of h.

▶ *Exercise* A4.4.2   Step through the algebra of the Simpson integral formula derivation. □

In equation (A4.28) we have cheated a little, because, for a given step size h we have to evaluate the function at intervals of $h/2$, rather than h. To acknowledge this and to cast our result into a standard form, we let $a = x_0 - h/2$, $h/2 \rightarrow h$, to obtain Simpson's formula for the integral from $x = a$ to $x = a + 2h$

$$\int dx\, y(x) \approx h[y(a) + 4y(a + h) + y(a + 2h)]/3. \quad (A4.29)$$

The error estimate for this formula is $-h^5 y^{iv}(a + h)/90$.

The composite Simpson formula for an *even* number of intervals n is readily obtained from the last equation. For the interval from $x = a$ to $x = a + nh$ the Simpson rule is

$$\int dx\, y(x) \approx h[y(a) + 4y(a + h)y(a + nh) + 4\Sigma y(a + (2j + 1)h) + 2\Sigma y(a + 2jh) \quad (A4.30)$$

In this formula the sum runs from $j = 1$ to $j = n/2 - 1$.

The error estimate for the Simpson formula is $-nh^5 y^{iv}/90$, where the derivative is an upper limit in the range a to $a + nh$. The restriction of n to an even number of intervals is sometimes troublesome in applying the Simpson formula, for example in integrating experimental data.

Comparing trapezoid and Simpson formulas

In comparison to the trapezoid formula, Simpson's formula can be very much more accurate. For example, any function that can be described by a parabolic or cubic equation in the range of integration will be integrated exactly, apart from round-off errors. Furthermore, in the Simpson formula the accuracy/time factor is 16, a factor of 4 larger than that for the trapezoid rule.

Numerical comparison of the two numerical integration formulas is suggested in Problem [A4.5] at the end of this chapter, and in L5 where determining electrostatic potentials by numerical integration is presented.

The problems of integration over an infinite range and of integrating functions that change rapidly in a small region are considered in a practical manner in the numerical-analysis textbook by Acton.

Integration formulas based on truncated Taylor series for functions tabulated at equal intervals, such as those that we derived here, are called *Newton-Cotes* formulas. Those for higher-derivative approximations are available in numerical-analysis handbooks, such as Abramowitz and Stegun. Such integration formulas are most useful when the integrand is time-consuming to obtain.

One alternative method of integration is to find optimal points in the interval of integration in order to minimize the integration error, then to evaluate the integrand at these special points. Such *Gauss-Lagrange* methods are used in applications where the evaluation of the function is difficult or expensive, so that minimizing the number of points is a prime consideration for efficiency. When integrating experimental data, it is often most convenient, and sometimes necessary, to specify the points to be used. In particular, equally spaced data points are often the most practical choice, so that Gauss-Lagrange methods are then impractical.

A practical method, suitable for both numerical integration and differentiation, is to use a straightforward, low-order formula with successively smaller values of h, $h/2$, etc, then carefully to extrapolate to the limit $h \rightarrow 0$ by an algebraic method. This is called *Romberg's method*, and it can be very effective. It is discussed in detail in Vandergraft's text on numerical computations.

A4.5 {1}   Diversion: Analytic evaluation by computer

As contrasted with numerical evaluations emphasized in A4.1 through A4.4, computers may also be used to help in the analytic evaluation of expressions. For example, in analytical problems involving much tedious algebra and many cancellations of terms among long expressions, human calculation is very error prone and inefficient. However, by programming the rules of algebra for the manipulation of symbols, this work can be done exactly by computer manipulation. In such algebraic manipulations, the variables are groups ("strings") of characters, rather than numbers that are used in numeric programming.

The use of computer programs for analytic evaluation has revealed that many of the standard tables of integrals have error rates of 5% or more. See, for example, the study of errors that is discussed in the book by Klerer and

Grossman. The errors occur predominantly in incorrect specification of the parameter range over which the integration formulas are valid.

In higher-level applications, for example to research in mathematics, celestial mechanics, gravitational and quantum physics, variables may not commute, or complicated derivatives may be required. Human frailty is then better supplemented by computer predictability. The research areas of general relativity, where the Ricci tensor calculus is used extensively, and quantum electrodynamics and chromodynamics (requiring the Dirac-Feynman algebra) make frequent use of analytic evaluation by computer.

▶ *Exercise* A4.5.1   If your computer system has good facilities for handling character strings, write a small program segment that will factor the symbolic expression $a*b + a*c$ into $a*(b + c)$. For more generality, include a facility to recognize that $a*b = b*a$, then use it to factor, for example, $b*a + a*c$. □

An interesting article that provides an introduction to algebra by computers is that by Pavelle, Rothstein, and Fitch in *Scientific American*.

A4.6 {2}   Curve fitting by splines

The derivative (studied in A4.3) measures an intensive property of a curve, its slope at a point. The integral (in A4.4) measures an extensive property, the area under a curve in some region. The prediction of values between points is the subject of *interpolation*. One method of curve fitting and interpolation that also allows estimation of derivatives and integrals is spline fitting.

A spline often gives a representation of a function that is very close to an intuitive description. The drafting instrument called a spline may be familiar to engineers from its use in the fairing and lofting of surfaces, as in designing to reduce drag on hulls and airfoils. Computerized curve fitting by splines therefore has important applications in computer-assisted design and manufacturing, CAD/CAM.

This section concentrates on motivating the use of splines and developing formulas for cubic splines. Computing laboratory L7 has Pascal and Fortran programs for spline fitting and extensive numerical exercises in which you can explore spline fitting.

Introduction to splines

Suppose that we are given n values of a function $y_j = y(x_j)$. For simplicity, we assume that successive arguments x_j each differ by an amount h, starting at x_1. Thus, $x_j = x_1 + (j - 1)h, j = 2,3,\ldots,n$. Typically, we want to interpolate, to find derivatives, and to evaluate integrals of $y(x)$ in the range x_1 to x_n.

One traditional method, using a truncated Taylor series to approximate $y(x)$, is to fit a single polynomial of order $n - 1$ through the n points, then to estimate the desired quantities from the polynomial. Although this method is hallowed by tradition (see for example Abramowitz and Stegun or Christman, Volume I, Chapter 4), it has the severe disadvantage that as n increases the polynomial fit will generally have many wiggles in it, especially if the y_j arise from an approximate numerical procedure or are experimental data.

▶ *Exercise* A4.6.1   Show that a polynomial of order n may have as many as $n - 1$ changes of direction. An extensive discussion of the polynomial wiggle problem and examples of it are given in Chapter 5 of Maron's text. □

However, we may know that such a wiggly behavior is inconsistent with the problem at hand. Can we find a curve-fitting method that guarantees a smooth fit through a sequence of points?

The cubic-spline fitting method satisfies this requirement. The spline method is the numerical analog of using a thin, uniform, flexible strip (a drafting spline[1]) to draw smooth curves through and between points on a graph. These points through which the curve is required to pass are called the *knots* of the spline. A sequence of data points and a cubic spline fit through them are sketched in Figure A4.6.1. If the curvature of the fit is to be smooth, but we are not too concerned about high-order derivatives of $y(x)$, then the function could be approximated over limited intervals by a cubic, $s(x)$, having the following properties:

(1) Within each subinterval $x_j \leqslant x \leqslant x_{j+1}$ $s(x)$ is a cubic polynomial.
(2) Each of s, s', s'' is continuous over the whole range $x_1 \leqslant x \leqslant x_n$.
(3) At each knot $x = x_j$, the spline fit goes through the data, that is $s(x_j) = y_j$, $j = 1,2,...,n$.

Note that $s(x)$ is a *different* cubic in each interval, so that there are $n - 1$ cubics describing the spline fit through n data points.

Derivation of spline formulas

Given these conditions on a cubic spline fit, we now proceed to derive an appropriate algorithm for spline fits. Since the fitting conditions relate to derivatives, we make Taylor expansions about x_j for x in the range x_j to x_{j+1} and for each interval $j = 1$ to $j = n$. For a cubic polynomial, derivatives past the third are zero, and we can immediately calculate the necessary third derivative by noting that a cubic has a constant third derivative, given in terms of second derivatives at the knots by

$$s'''_j = (s''_{j+1} - s''_j)/h. \tag{A4.31}$$

---

[1]The drafting spline and the gear spline in mechanical engineering are related in name only.

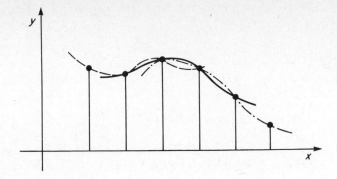

Figure A4.6.1   Schematic of a spline fit with three cubics to a series of data points. The cubics in each interval are shown beyond their range of use, thus indicating their matching at the knots.

Thus, the spline $s(x)$ can be written, using $s_j = y_j$ [according to condition (3)],

$$s(x) = y_j + s'_j(x - x_j) + s''_j(x - x_j)^2/2 + (s''_{j+1} - s''_j)(x - x_j)^3/6h \quad (A4.32)$$

Our goal is to solve for the derivatives in terms of the y_j by use of the spline continuity conditions (2). By differentiating equation (A4.32), we obtain

$$s'(x) = s'_j + s''_j(x - x_j) + (s''_{j+1} - s''_j)(x - x_j)^2/2h, \quad (A4.33)$$

and, on differentiating this equation,

$$s''(x) = s''_j + (s''_{j+1} - s''_j)(x - x_j)/h. \quad (A4.34)$$

We can relate the first derivatives at the knots by using (A4.33) with j replaced by $j - 1$ then $x \rightarrow x_j$, to find that

$$s'_j - s'_{j-1} = (s''_{j-1} + s''_j)h/2. \quad (A4.35)$$

From (A4.32) with $x = x_{j+1}$ we have, using condition (3),

$$y_{j+1} = y_j + s'_jh + s''_jh^2/2 + (s''_{j+1} - s''_j)h^2/6. \quad (A4.36)$$

From (A4.32) with j replaced by $j - 1$, then $x \rightarrow x_j$, and condition (3), we have

$$y_j = y_{j-1} + s'_{j-1}h + s''_{j-1}h^2/2 + (s''_j - s''_{j-1})h^2/6. \quad (A4.37)$$

Subtract the second equation from the first, substitute (A4.35), and rearrange to obtain for $j = 2,3,\ldots,n - 1$,

$$s''_{j-1} + 4s''_j + s''_{j+1} = d_j, \qquad \text{(A4.38)}$$

where

$$d_j = 6[(y_{j+1} - y_j - (y_j - y_{j-1})]/h^2. \qquad \text{(A4.39)}$$

The set of linear equations (A4.38) can be solved if the derivatives s''_1 and s''_n are given as boundary conditions. Problem [A4.7] shows that the spline fit is optimal if these derivatives are zero. This condition is equivalent to the analogous drafting spline having unconstrained ends; hence the term *natural spline* for a spline fit with the boundary conditions of zero second derivatives at the endpoints.

The equations (A4.38) are solved iteratively by first eliminating s''_{j-1} between pairs of these equations. Then we set

$$a_2 = 4, \; b_2 = d_2 - s''_1, \qquad \text{(A4.40)}$$

$$a_j = 4 - 1/a_{j-1}, \; b_j = d_j - b_{j-1}/a_{j-1}, \; j = 3,4,\ldots,n - 1. \qquad \text{(A4.41)}$$

The spline equations (A4.38) then become

$$a_j s''_j + s''_{j+1} = b_j, \; j = 2,3,\ldots,n - 1. \qquad \text{(A4.42)}$$

From this, starting at $j = n - 1$ and working downward to smaller j, we find the spline second derivatives at the knots

$$s''_{n-1} = b_{n-1}/a_{n-1}, \; s''_j = (b_j - s_{j+1})/a_j, \; j = n - 2,\ldots,2. \qquad \text{(A4.43)}$$

▶ *Exercise* A4.6.2   Verify all the steps in the spline derivation. (Spare the spline and spoil the child.) ☐

The method of solving the linear equations (A4.38) for the spline derivatives s''_j is equivalent to Gaussian elimination applied to inverting a tri-diagonal matrix. This connection is used in more general treatments of spline fitting, such as those in the monographs by De Boor and by Schumaker.

Algorithm for spline fitting

We summarize the algorithm for a cubic spline fit to n points $y_1, y_2, \ldots, y_n$ as follows:

(1) Compute the coefficients a_j according to the iteration scheme

$$a_2 = 4, \; a_j = 4 - 1/a_{j-1}, \; j = 3,4,\ldots,n - 1. \tag{A4.44}$$

These spline coefficients depend only on the number of data points and not on the data values.

(2) With j increasing, and for the given data values, compute the first differences

$$e_j = y_j - y_{j-1}, \; j = 2,3,\ldots,n. \tag{A4.45}$$

Thence compute the second differences

$$d_2 = 6(e_3 - e_2)/h^2 - s''_1, \tag{A4.46}$$

$$d_j = 6(e_{j+1} - e_j)/h^2, \; j = 3,4,\ldots,n - 2, \tag{A4.47}$$

$$d_{n-1} = 6(e_n - e_{n-1})/h^2 - s''_n, \tag{A4.48}$$

and the coefficients

$$b_2 = d_2, \; b_j = d_j - b_{j-1}/a_{j-1}, \; j = 3,4,\ldots,n - 1. \tag{A4.49}$$

Note that the second derivatives at the boundaries s''_1, s''_n have to be supplied. They are zero if the natural cubic spline fit is made.

(3) With j decreasing, compute the second derivatives at the spline knots

$$s''_{n-1} = b_{n-1}/a_{n-1}, \tag{A4.50}$$

$$s''_j = (b_j - s''_{j+1})/a_j, \; j = n - 2,\ldots,2. \tag{A4.51}$$

(4) With j increasing, compute the first derivatives at the knots

$$s'_1 = e_2/h - s''_1 h/3 - s''_2 h/6, \tag{A4.52}$$

$$s'_j = s'_{j-1} + (s''_{j-1} + s''_j)h/2, \; j = 2,\ldots,n, \tag{A4.53}$$

and the third derivatives at the same points

$$s'''_1 = 0, \; s'''_n = 0, \tag{A4.54}$$

$$s'''_j = (s''_{j+1} - s''_j)/h, \; j = 2,\ldots,n - 1. \tag{A4.55}$$

(5) To find values and derivatives at any point x in the range x_1 to x_n, locate the appropriate interpolation index i by

$$i = \text{trunc}[(x - x_1)/h] + 1. \tag{A4.56}$$

Then set $e = x - x_1 - (i - 1)h$. We obtain the interpolated value

$$s(x) = y_i + (s'_i + (s''_i/2 + s'''_i e/6)e)e, \tag{A4.57}$$

the interpolated first derivative

$$s'(x) = s'_i + (s''_i + s'''_i e/2)e, \tag{A4.58}$$

and the interpolated second derivative

$$s''(x) = s''_i + s'''_i e. \tag{A4.59}$$

The third derivative for the cubic spline is constant in the ith interval, so it is simply

$$s'''(x) = s'''_i, \tag{A4.60}$$

in which (A4.55) is used for the right-hand side.

Spline properties

Several properties of cubic-spline fits are useful in numerical analysis and applied science. For example, the integral of $s(x)$ over any range between x_1 and x_n can be obtained by summing the integrals in each subrange, using formula (A4.32) for $s(x)$. The details of this integration procedure form an exercise in L7.

If the endpoint conditions on the second-derivatives s''_1, s''_n are not known, and natural spline conditions (second derivatives zero) are used, be careful about using values of n that are small, say $n < 4$, or x values very close to the endpoints x_1 and x_n, because the use of the natural spline condition at the endpoints will unduly influence the values obtained.

Computing laboratory L7 has many examples on properties of splines, including the effects of boundary conditions and the use of cubic splines for interpolation, differentiation, and integration.

Development of splines and computers

As an interlude in this extensive chapter on numerical analysis, we sketch how the scientific applications of curve fitting by splines have been closely related to the development of computers.

The spline-fitting prescription involving steps (1) through (5) requires a computing device with significant random-access memory, because for an n-point cubic spline about $9n$ points are computed, and more than half these must be available after the fourth step. Unlike polynomial interpolation formulas (see, for example, Abramowitz and Stegun), which have coefficients depending only on the order of the polynomial assumed, in spline fitting steps (2) through (5) must be repeated whenever even a single function value y_j is changed. In hand calculation of a fitting curve by splines, the large number of arithmetic operations and the strong dependence on the function values is a hindrance, especially because any error propagates throughout the calculation in a complicated way. Therefore, spline methods were seldom used until computers and pocket calculators with extensive memory were widely available. Interpolation by drawing with drafting splines is of great antiquity; one has to draw the line somewhere.

Historically, splines received greatest attention from mathematicians, until about 1960 when the digital computer was developed into a practical, common machine. As a consequence of the attention from mathematicians, the mathematical formalism of splines is very well developed, as exemplified in the monographs by Schultz and by Schumaker. On the other hand, the practical applications of splines are still under extensive development in the applied sciences. The book by De Boor, which contains extensive Fortran programs, has been influential in this development.

By contrast with splines, Fourier expansion techniques (as developed in A5) started in the early 1800s in applicable mathematics, especially in the theory of heat conduction. Only later were Fourier expansion methods developed in mathematical rigor and generality. From these two examples we see that pure mathematics, computing machines, and applied science form a research triangle, each contributing to the development of the others.

A4.7 {2} The least-squares principle

In the preceding sections we considered derivatives, integrals, and curve fitting of functions such that at each $x = x_j$ the fitted values agreed exactly with the given $y(x_j) = y_j$. However, if the y_j are data from observations, then they will have associated uncertainties (errors) that should influence the fit. The most realistic fitted curve to the whole set of data may not necessarily pass through each y_j. For example, if we expect that the data points should lie on a straight line, then the fit shown in Figure A4.7.1 may be acceptable.

Suppose that data point j has error estimate e_j. These errors are assumed to arise from random effects and to be such that after many observations the probability of obtaining a value between $y_j - e_j$ and $y_j + e_j$ is the same for each data point. Usually, a probability value of about 2/3 is chosen; thus about 1/3 of the data points in a large sample are expected to lie outside this range, either above it or below it.

Consider the quantity chi-squared, defined here by

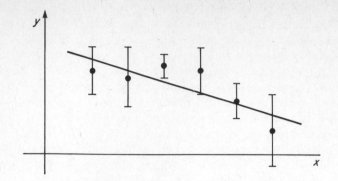

Figure A4.7.1   Data with errors in the abscissa, and a straight-line least-squares fit to these data.

$$\chi^2 = \Sigma[y_j - \phi(x_j)]^2\omega_j, \tag{A4.61}$$

in which the summation is over the data points from $j = 1$ to $j = n$. The weight factor ω_j is usually related to the error estimate by $\omega_j = 1/e_j^2$. The theory of probability shows that if the fitting function ϕ adequately describes the data y_j, then the probability that χ^2 has a value of n or less is greater than 1/2. (See, for example, Chapter 4.3 of Pizer's text on numerical computing, or Chapter 5 of Ehrlich.) Although more refined confidence limits, depending on χ^2 and n, can be estimated, this rule of thumb is usually adequate in real-world situations, in which the randomness of the errors is not necessarily a reasonable assumption and the error estimates e_j are usually quite ill-determined.

The minimization of a sum of squares of differences between data and fitting function is called a *least-squares fit*. Historically, the method of least squares was developed by Gauss and Legendre in the early nineteenth century. Both Gauss and Legendre were trying to determine astronomical orbits accurately from sequences of imprecise observations, which were usually greater in number than the number of unknowns for the orbit. Gauss also related the least-squares principle to the theory of probability. The discovery of the method and the controversy over the credit for its invention are discussed by Goldstine.

Choice of least squares

The criterion for using squares in the definition (A4.61) is worth discussion. Why not minimize instead just the sum over the n data points of the weighted differences $\{y_j - \phi(x_j)\}\omega_j$? This has a minimum, equal to zero, for any function ϕ satisfying $\Sigma \phi_j\omega_j = \Sigma y_j\omega_j$. In particular, the value $\phi_j = y_{av}$, where y_{av} is given by

$$y_{av} = \Sigma y_j\omega_j/\Sigma \omega_j \tag{A4.62}$$

is a possible choice. However, this choice of the best fit is not very realistic. For example, an oscillating function is not very well approximated by its average.

As an alternative to simply calculating averages, suppose that the sum of the absolute values of weighted differences were minimized. Unfortunately, this is fraught with difficult mathematical analysis, especially since the absolute value is singular near zero, which is the interesting point that we would like to reach.

▶ *Exercise* A4.7.1   Show that the function $|x|$ has slope $+1$ for $x > 0$, and slope -1 for $x < 0$. Then show that the absolute value function is not well behaved near the origin. □

The choice of squares in (A4.61) provides the least-complicated, nontrivial, tractable, objective, fitting criterion. One disadvantage of a least-squares fitting criterion is that χ^2 gets at least 4 times as much contribution from points at which the data and fitting function differ by more than twice the error estimate e_j as from points where they agree to within the error estimate. Thus the assignment of relative errors is very important in least-squares fitting. Such a fit may not agree very well with the best eyeball fit, which probably corresponds roughly to minimizing the absolute values of the differences between data and fitting function. However, being subjective, eyeball fits depend sensitively on the experience and desires of the fitter.

In the rest of this chapter we derive general properties and a few widely used special results in least-squares fitting. Particular applications are made in deriving Fourier expansion coefficients in A5 and in computing laboratory L8 on least-squares analysis of data.

Derivation of least-squares equations

We now consider a fairly formal derivation of the least-squares equations. If you will persist with this formalism, you will find that you have a general framework within which to construct a variety of least-squares analyses, including fits to polynomials, or to straight lines, or to trigonometric (or complex exponential) functions—the subject of Fourier expansions in A5.

In general, given a choice of the fitting function ϕ, the parameters of this function that minimize χ^2 can only be determined by intelligent application of cut-and-try methods. However, in many applications the parameters to be determined appear linearly in the definition of ϕ;

$$\phi(x) = \Sigma a_l \phi_l(x), \ l = 1,2,\ldots,N, \tag{A4.63}$$

in which a set of N functions has been assumed. In this expansion it is assumed that the functions $\phi_l(x)$ do not contain parameters to be determined.

We call the process by which the a_l in equation (A4.63) are adjusted so as to minimize χ^2 a *linear-least-squares fit*. This is to be distinguished from the more restrictive least-squares fit to a straight line, $\phi_1(x) = 1$, $\phi_2(x) = x$ (so that $\phi(x) = a_1 + a_2 x$), commonly also called a linear least-squares fit, instead of a straight-line least squares fit, as we denote it.

Finding the linear-least-squares coefficients a_l, $l = 1, 2, \ldots, N$, that minimize χ^2 requires that the derivative of χ^2 with respect to each of the a_l be zero. Consider any one of the coefficients, say a_k. Take the derivative on both sides of (A4.61) with respect to a_k, then set the result to zero to find a condition for a minimum of χ^2. We readily find that

$$0 = \Sigma \, [y_j - \phi(x_j)]\phi_k(x_j)\omega_j. \tag{A4.64}$$

Here the sum is over all data points j from $j = 1$ to $j = n$. By inserting the expansion (A4.63) in this equation we get

$$\Sigma a_k \Sigma \phi_k(x_j)\phi_l(x_j)\omega_j = \Sigma y_j \phi_l(x_j)\omega_j, \quad l = 1, 2, \ldots, N. \tag{A4.65}$$

This is a set of N linear equations for the coefficients a_k that minimize χ^2. In general, these equations have to be solved by matrix methods.

▶ *Exercise* A4.7.2   Derive equation (A4.65), starting from the least-squares condition on χ^2, (A4.64), and taking the indicated derivatives with respect to the fitting parameters a_k. □

Use of orthogonal functions

The general linear-least-squares equations (A4.65) serve as a starting point for least-squares fits using common functions. Of greatest practical importance are *orthogonal functions*. These are sets of functions such that with weight factor ω_k and summation over the observation points x_j, and two functions ϕ_l and ϕ_k in the set are constrained by

$$\Sigma \, \phi_k(x_j)\phi_l(x_j)\omega_j = 0, \, l \neq k. \tag{A4.66}$$

For example, we will see in A5 that complex-exponential functions can be made orthogonal if the x_j are appropriately chosen in the range 0 to 2π. Indeed, this property is exploited in Fourier expansions in A5, L9, and L10.

▶ *Exercise* A4.7.3   Suppose that $\mathbf{l}$ and $\mathbf{k}$ are two three-dimensional vectors with components expressed in Cartesian coordinates.
(*a*) Write down, in a form analogous to (A4.66), the orthogonality condition for these two vectors.
(*b*) If $\mathbf{k}$ is a fixed vector, what is the set of all vectors $\mathbf{l}$ that are orthogonal to $\mathbf{k}$? □

If the orthogonality property (A4.66) is satisfied, then (A4.65) can be solved immediately because the sum on the left-hand side collapses to a single nonzero term, resulting in

$$a_k = \Sigma y_j \phi_k(x_j)\omega_j / \Sigma \phi_k^2(x_j)\omega_j, \qquad (A4.67)$$

in which both sums are over the n data points. This is the formula for the coefficients in a linear-least-squares fit with orthogonal functions. If the weights ω_j are determined from error bars on the data, suitable orthogonal functions will usually not be found. A common compromise is to use this equation with all weights equal to unity for orthogonal functions, and to use $\omega_j = 1/e_j^2$ in (A4.61) for other fitting functions.

Another advantage of using orthogonal functions in linear-least-squares fits is that each fitting coefficient a_k is obtained independently of the others, as (A4.67) shows directly. Thus, if the set of functions is enlarged by increasing N, the previously computed coefficients do not need to be redetermined. This practical property does not hold for non-orthogonal functions, because the coefficients are linked together by (A4.65).

The independence of the coefficients in a linear-least-squares fit with orthogonal functions was a great advantage in the days of hand calculations, because errors made in computing one coefficient did not necessarily ruin the other coefficients. With fast, reliable computers, and fairly reliable numerical analysts and programmers, this is now less of a consideration. By contrast with the use of orthogonal functions in least-squares fitting, the spline analysis in A4.6 links together all the fitting functions.

Fitting averages and straight lines

Here we derive and discuss the simplest least-squares fits, those to a constant and to a straight line. We begin with the more-general consideration of polynomials as the fitting function and the powers of the independent variable multiplying the coefficients:

$$\phi_k(x) = x^{k-1}, \quad \phi(x) = \Sigma a_k x^{k-1}, \qquad (A4.68)$$

in which the sum is over the power $k = 1$ to $k = N$. ϕ is thus a polynomial of order $N - 1$, with N coefficients. The powers of x do not form a set of orthogonal polynomials, although such can be constructed from linear combinations of the powers, as Problem [A4.9] shows. Eschewing this choice, we must use the general linear equations (A4.65). However, for small N these equations can be written down and solved directly.

The fit producing the average value has $N = 1$ and thus $\phi_1 = x^o = 1$. Directly from (A4.65) we have the least-squares estimate of the average value

$$a_1 = \Sigma y_j \omega_j / \Sigma \omega_j, \qquad (A4.69)$$

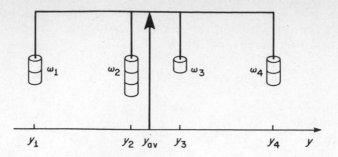

Figure A4.7.2   Mechanical analog of equation (A4.69). The average value is located at the balance point for the weights $\omega_1 = 2$, $\omega_2 = 3$, $\omega_3 = 1$, $\omega_4 = 2$.

with the sum over all the data points. This is just the weighted average of the n data values y_j, each taken with weight factor ω_j. The analogy with statics is that the ω_j are the weights (literally) and the y_j are their distances from the origin. This is illustrated in Figure A4.7.2, in which the average over the $N = 4$ data points is analogous to the fulcrum position (balance point) y_{av} for weights ω_j hanging at distances y_j from the origin.

Fitting to a straight line by least squares is an important topic, which in some areas of applied science is also called "linear regression" analysis. For the straight-line least-squares fit $N = 2$, $\phi(x) = a_1 + a_2 x$, and (A4.65) reduces to the pair of equations

$$a_1 S_0 + a_2 S_1 = Y_0, \tag{A4.70}$$

$$a_1 S_1 + a_2 S_2 = Y_1, \tag{A4.71}$$

where, for $p = 0,1,2$,

$$S_p = \Sigma x_j^p \omega_j, \tag{A4.72}$$

$$Y_p = \Sigma y_j u_j^p \omega_j. \tag{A4.73}$$

In both these expressions the summations are over the data point index j. Thus, the least-squares fit to a straight line has intercept given by

$$a_1 = (Y_0 S_2 - Y_1 S_1)/\Delta, \tag{A4.74}$$

and slope given by

$$a_2 = (Y_1 S_0 - Y_0 S_1)/\Delta, \tag{A4.75}$$

where

$$\Delta = S_0 S_2 - S_1^2. \tag{A4.76}$$

▶ *Exercise* A4.7.4   Derive formulas (A4.74), (A4.75), and (A4.76) for the straight-line least-squares fit. □

The formulas given for the straight-line least-squares fit can be numerically unstable because of subtractive cancellations. For example, if the observations are clustered around large values of x, then each of the sums may be relatively large but Δ may be quite small.

An investigation of the numerical stability of the straight-line least-squares fit formulas is given as Problem [A4.10].

▶ *Exercise* A4.7.5   Show that rearrangement of (A4.72) through (A4.76) can be made to produce a numerically more stable form of the straight-line least-squares formulas. Namely, for the intercept

$$a_1 = y_{av} - a_2 x_{av}, \tag{A4.77}$$

in terms of the slope

$$a_2 = S_{xy}/S_{xx}. \tag{A4.78}$$

In these equations there appear the weighted averages of x and y values, as well as the sums

$$S_{xy} = \Sigma (x_j - x_{av})(y_j - y_{av})\omega_j, \tag{A4.79}$$

$$S_{xx} = \Sigma (x_j - x_{av})^2 \omega_j. \tag{A4.80}$$

□

In the straight-line least-squares fit formulas we have used an expansion in polynomials that are not orthogonal, and we see that the coefficients a_1 and a_2 are not obtained independently. Orthogonal polynomials for use in least-squares fitting can be systematically generated by a procedure derived in Problem [A4.9]. Recall that the use of orthogonal polynomials has the practical advantage that one can readily change from linear, to quadratic, to higher-order orthogonal polynomials without changing the lower-order coefficients, or vice versa.

Computing laboratory L8 emphasizes straight-line least-squares fitting. The modification of the analysis to allow for errors in the x variable as well as in the y variable is considered in L8.1. From the Pascal or Fortran programs developed in L8.2 you can explore least-squares analysis of data, interesting samples of which are provided in L8.3.

Problems in numerical analysis

[A4.1] {2}   (Round-off and truncation noise) This problem investigates how round-off and truncation noise propagate when many arithmetic operations are performed.

(a) For round-off, show by explicit examples, or by describing the build-up of round-off effects as a random-walk problem, as considered in L6.1, or by using a probabilistic model of round-off error propagation (see Hamming), that the round-off noise after n operations between numbers of similar size (ignoring subtractive cancellation) grows roughly proportionally to $\sqrt{n}$.

(b) For truncation, show by examples, or by arguing from the binomial approximation in A3.5, that the truncation noise after n operations between numbers of similar size, ignoring subtractive cancellation, grows roughly proportionally to n.

(c) If you have a pocket calculator, check whether it uses round-off or whether it uses truncation. To do this, repetitively apply two arithmetically inexact inverse operations n times, where n values of, say, 16, 64, 256, and 1024 are used. For example, calculate $\sqrt{2}$ then square the result, repeating the square root and squaring operations n times in succession applied to the same variable. Does the error grow roughly proportionally to $\sqrt{n}$ (for round-off) or to n (for truncation)?

[A4.2] {1}   (Numerical convergence of series) The geometric series studied in A3.2 has the closed form $S_n = a(1 - r^n)/(1 - r)$, and converges to $S = a/(1 - r)$ if $|r| < 1$. To investigate the hazards of computer arithmetic, especially the effects of subtractive cancellation and round-off, let $r = -1 + e$, where e is a small positive number. Use $a = 1$, so that, analytically at least, $S = 1/(2 - e)$.

(a) Write and run a program which uses about 7 significant figures at each step to calculate S_n by direct summation of the geometric series. Assume that $a = 1$, and input values of e, for example, $e = 0.1, 0.01, \ldots, 10^{-4}$. Compare the summed result with the closed forms for S_n and S in order to see the effects of numerical noise.

(b) Suppose that $e = 10^{-8}$. Use the binomial theorem to estimate how many terms would be needed to get about 50% accuracy in S. Do not attempt to compute this S_n by direct summation. Why not?

[A4.3] {2}   (Subtractive cancellation in quadratics) In solving quadratic equations with $b^2 \gg |4ac|$ we found in A4.2 that the small root is numerically unstable because of subtractive cancellation.

(a) As one way of avoiding this problem, use a binomial approximation to show that $\sqrt{(b^2 - 4ac)} \approx b - 2ac/b$, so that the small root is approximated by $-c/b$, and the larger root by $-b/a$.

(*b*) From these two roots we have that $(x + c/b)(x + b/a) \approx 0$. How is this related to the original quadratic and to the approximations made in solving it?

(*c*) Show that the solution of the quadratic equation can be written exactly as $x = -2c/[b \pm \sqrt{(b^2 - 4ac)}]$ and thus that for $b > 0$ subtractive cancellation does not occur for the smaller root.

(*d*) Write a program that gives the real roots of the quadratic equation, given as input a, b, c. The algorithm used should be insensitive to subtractive cancellation for either sign of b. Therefore either the conventional form of the formula for the root or the form derived in part (*c*) should be used, as appropriate.

[A4.4] {1} (Comparison of polynomial algorithms) Consider the Maclaurin function expansion to n terms (ignoring the remainder) $\tan^{-1}(x) = x\Sigma(-x^2)^k/(2k + 1)$, in which the sum runs from $k = 0$ to $k = n - 1$.

(*a*) Write and run a program that compares the direct and Horner polynomial algorithms for the tan^{-1} function over the range $0 \leqslant x \leqslant \pi/4$.

(*b*) Determine an appropriate n that gives an accuracy of the same order as the numerical noise for your calculating engine.

(*c*) Which polynomial algorithm, direct or Horner, gives more accurate results for your calculator?

[A4.5] {2} (Numerical integration algorithms) In numerical integration a good method for testing accuracy and comparing integration algorithms is to choose a function that can be integrated analytically over the desired range and that resembles functions to which the numerical integration algorithm may be applied. For example, the functions to be integrated might be approximately sinusoidal. Therefore consider $y(x) = \sin(x)$ and the range of integration $a = 0$ to $a + nh = \pi$. The comparison value for the integral is then 2.

(*a*) Use the composite trapezoid rule (A4.27) and a computer or pocket calculator program to estimate numerically $\int dx \sin(x)$. Investigate the accuracy as a function of the step size h. Since the second derivative y'' can be determined analytically, check how closely the error estimate given agrees with the actual error.

(*b*) Make a similar comparison using the composite Simpson formula (A4.30). Check the error estimate given there by using y^{iv} obtained analytically. Note that h must be chosen so that n, the number of intervals in 0 through π, is an even integer.

(*c*) Assume that the fractional accuracy required in the integral is one part per million, which is (it is hoped) larger than the numerical noise in your computing system. For this integral, investigate numerically how much more efficient in accuracy/time the Simpson formula is than the trapezoid formula.

[A4.6] **{2}**   (Integral approximations to series) Here we consider a method of approximating the summation of series by integration. The trapezoid formula (A4.27) can be rewritten to approximate the sum by the integral.
(*a*) Show thereby that

$$\Sigma\, y(a\, +\, kh) \approx \int dx\, y(x)/h\, +\, [y(a)\, +\, (y(a\, +\, nh)]/2, \qquad \text{(A4.81)}$$

where the sum over k is from 0 to n, and the integral is from a to $a + nh$. Hence show that an approximation to the partial sum $S_n = \Sigma\, a_k$, where the a_k are the terms in the series, is obtained by the choice of $h = 1$; namely $S_n \approx (a_1 + a_{n+1})/2 + \int dx\, a(x)$.
(*b*) For the geometric series (A3.4) show that this method results (for $0 < r < 1$) in the estimate $S \approx -1/[r \ln (r)] + r/2$, where S is the limit as $n \rightarrow \infty$ of S_n, whereas the exact result is $S = 1/(1 - r) - 1 = r/(1 - r)$. Show that the method is poor in this example because a mesh size $h = 1$ is used.

The approximation of a series by an integral is most suitable for slowly converging, alternating series.

[A4.7] **{3}**   (Natural splines have minimal curvature) The natural spline provides a fit with minimum curvature. To show this, prove that a natural spline fit minimizes the quantity $C = \int dx\, [s''(x)]^2$, which is related to the average curvature of the spline fit, provided that the natural-spline endpoint conditions $s''_1 = s''_n = 0$ are satisfied. *Hint*: Show that if g is any other spline fit with different endpoint conditions, then

$$\int dx\, [g''(x)]^2\, =\, \int dx\, [s''(x)]^2\, +\, \int dx\, [g''(x)\, -\, s''(x)]^2, \qquad \text{(A4.82)}$$

which necessarily exceeds C, which is the first integral on the right-hand side. Compute the second integral on the right-hand side by expanding the square and then splitting the integral into a sum of integrals over segments of length h, integrating by parts, resumming, and finally using the natural endpoint conditions.

[A4.8] **{3}**   (Program for cubic splines) Write an interactive cubic spline fitting program that inputs n, h, the data points y_j, $j = 1,...,n$, and the second derivatives at the endpoints, then uses the algorithm derived in A4.6 to calculate the necessary derivatives at the knots and store them. Then the program should accept a range of x values and provide the values and derivatives according to step (5). The sample Pascal or Fortran programs given in computing laboratory L7 will provide useful starting points.

[A4.9] **{3}**   (Orthogonal polynomials for least-squares fit) For least-squares fits with polynomials, it is advantageous that the polynomials be orthogonal, as discussed in A4.7. Prove the orthogonality of the polynomials generated by the following recurrence relations:

$$\phi_0(x) = 1 \tag{A4.83}$$

$$\phi_{i+1}(x) = (x - a_{i+1})\phi_k(x) - \beta_{i+1}\phi_{i-1}(x), \ i = 1, 2, \ldots,$$

where

$$a_{i+1} = \Sigma\, x_j[\phi_i(x_j)]^2\omega_j/\Sigma\, [\phi_i(x_j)]^2\omega_j, \tag{A4.84}$$

$$\beta_1 = 0, \ \beta_{i+1} = \Sigma[\phi_i(x_j)]^2\omega_j/\Sigma[\phi_{i-1}(x_j)]^2\omega_j \tag{A4.85}$$

In these equations the summations are over the data points x_j. Prove the orthogonality of the polynomials ϕ_i by the method of induction applied to $k = 0, 1, \ldots, i$, namely that the summation over the data points x_j

$$\Sigma\, \phi_{i+1}(x_j)\phi_k(x_j)\omega_j = 0, \tag{A4.86}$$

for $i \neq k$.

[A4.10] {2} (Stability of least-squares algorithms) To test the straight-line least-squares fitting formulas, write a program that compares the numerical stability of the two forms of the fit derived in A7.2 for the following data, combined with weight factors of unity:

x_k 9.990 9.995 10.000 10.005 10.010
y_k 5.990 5.998 6.000   6.002 6.006

Use about 6 significant figures in the arithmetic. A convenient starting program, in Pascal or Fortran, is available in L8.2. It uses the numerical method which Exercise A4.7.5 claims to be the more stable. Therefore only the coding for formulas (A4.77) through (A4.80) will be needed.

References on numerical analysis

Abramowitz, M., and I. A. Stegun, *Handbook of Mathematical Functions*, Dover, New York, 1964.

Acton, F. S., *Numerical Methods That Work*, Harper and Row, New York, 1970.

Christman, J. R., *Physics Programs for Programmable Calculators*, Vols. I and II, Wiley, New York, 1981 and 1982.

De Boor, C., *A Practical Guide to Splines*, Springer-Verlag, New York, 1978.

Ehrlich, R., *Physics and Computers*, Houghton Mifflin, Boston, 1973.

Gear, C. W., *Applications and Algorithms in Computer Science*, Science Research Associates, Chicago, 1978.

Goldstine, H. H., *A History of Numerical Analysis from the 16th through the 19th Century*, Springer, New York, 1977.

Gueron, M., "Enhanced Selectivity of Enzymes by Kinetic Proofreading," *American Scientist*, **66**, 202 (1978).

Hamming, R. W., *Numerical Methods for Scientists and Engineers*, McGraw-Hill, New York, 1962.

Klerer, M., and F. Grossman, *A New Table of Indefinite Integrals*, Dover, New York, 1971.

Maglich, B., "Discovery of Massive Neutral Vector Mesons," in *Adventures in Experimental Physics*, B. Maglich (ed.), **5**, 113 (1976).

Maron, M. J., *Numerical Analysis: A Practical Approach*, Macmillan, New York, 1982.

Pavelle, R., M. Rothstein, and J. Fitch, "Computer Algebra," *Scientific American*, December 1981, p.136.

Pizer, S., *Numerical Computing and Mathematical Analysis*, Science Research Associates, Chicago, 1975.

Schultz, M. H., *Spline Analysis*, Prentice-Hall, Englewood Cliffs, N.J., 1973.

Schumaker, L. L., *Spline Functions: Basic Theory*, Wiley, New York, 1981.

Vandergraft, J. S., *Introduction to Numerical Computations*, Academic Press, New York, 1978.

A5 FOURIER EXPANSIONS

In this chapter we develop mathematical and computational techniques applicable to describing data and functions in terms of orthogonal functions, particularly the cosine, sine, and exponential functions. In pure mathematics the term *Fourier analysis* often applies to expansions into arbitrary orthogonal functions. In the applied sciences the trigonometric functions are usually implied. Historically, the subject of Fourier analysis began with the researches of Joseph Fourier[1] into heat conduction through solids, made in the early nineteenth century.

The topics that we emphasize are the distinction between different kinds of Fourier expansions (A5.1) and the motivation for choosing among them, discrete Fourier transforms (A5.2), Fourier series and their applications (A5.3 through A5.6), Fourier integral transforms (A5.7), and numerical algorithms for fast Fourier transforms (A5.8). The applicable mathematics developed in this chapter has extensive applications in the computing laboratories; Fourier analysis of an EEG is a computing project in L9, and the analysis of atomic-clock resonance line widths by Fourier-integral transform and linear least-squares fitting techniques is researched in L10.

A5.1 {1} Overview of Fourier expansions

There are several motivations for using Fourier expansions in mathematics, engineering, and physics. Consider the least-squares fitting criterion in A4.7, with a function $\phi(x)$ used to describe a set of data y_j, $j = 1,2,...,n$. Given the linear-least-squares expansion

$$\phi(x) = \Sigma\, a_k \phi_k(x), \tag{A5.1}$$

in which the summation is over k from 1 to n, the coefficients a_k, which provide a least-squares fit of ϕ to the data, are given in terms of (A4.67)

$$a_k = \Sigma\, y_j \phi_k(x_j)\omega_j / \Sigma\, \phi_k(x_j)\omega_j, \tag{A5.2}$$

If the functions ϕ_ℓ are orthogonal to each other, that is, if

$$\Sigma\, \phi_\ell(x_j)\phi_k(x_j)\omega_j = 0,\, l \neq k, \tag{A5.3}$$

then the sum on the left side of (A4.65) reduces to a single term. Recall that in equation (A5.3) the summation is over the index j from 1 to n, and ω_j is the weight factor (usually related to the error) for data point j. The practicality of

[1]Fourier (1768–1830) was a French mathematical physicist and a science advisor to Napoleon I. For a readable biography of Fourier see the book by Herival.

the expansion (A5.1) is that, especially when the y_j comprise a set of data points, the functions $\phi_k(x)$ may be mathematically much more tractable, or scientifically more meaningful, than the y_j. This will aid in such operations as differentiation, integration, and interpolation, as considered in A4.3 through A4.6.

Transformations

The Fourier expansion coefficients a_k lead to an alternative representation, a transformation, of the information contained in the y_j. Transformations are already familiar to you. For example, in arithmetic when you use logarithms to facilitate multiplication or to allow a compressed scale on a graph, you are transforming numbers to logarithmic representation. (See, for example, the diversion on the logarithmic century in A6.4.) In geometry we often transform between Cartesian and polar coordinates to simplify the description. (See, for example, the programs in computing laboratory L2.)

In Fourier expansions when the variable x refers to a physical quantity that determines the values $y(x_j)$, then the expansion coefficients a_k comprise a set of complementary variables. For example, if the $\phi_k(x)$ are the circular functions when x denotes position, then the labels k indicate wavelength (or wave number) and each a_k gives the amount that wavelength component contributes to $y(x)$. If x denotes time, t, then the k refer to frequency, and each a_k gives that frequency's contribution to the signal amplitude y as a function of time.

Thus, we speak of Fourier transforming between the spatial domain and the wavelength domain, or between the time domain and the frequency domain. In computing laboratory L9, on the Fourier analysis of an EEG, the transformation is from time domain to frequency domain. In laboratory L10, where resonance line widths are analyzed, the first transformation is from frequency domain to time domain, followed by a logarithmic transformation to obtain a quantity with a linear dependence on time, and thus suitable for a straight-line least-squares analysis.

Nomenclature of Fourier expansions

The nomenclature of various forms of Fourier expansions is confusing and not always consistent among different references. Our nomenclature is summarized here and is related to subsequent sections.

The *discrete Fourier transform* uses summations in each of (A5.1), (A5.2), and (A5.3). It is most useful in the analysis of data (as discussed in section A5.2), especially because there is an algorithm, the fast Fourier transform (considered in A5.8 and L9.2), which makes computation of the discrete Fourier transform very efficient. Both the arguments x_j and the coefficients a_k form discrete sets, which is very convenient if the y_j are data.

The *Fourier series* uses the summation in (A5.1), but replaces sums by integrals in (A5.2) and (A5.3). Thus $y(x)$ should be available at arbitrary x, which is most suitable for analytical functions. However, the Fourier series can be represented by a discrete set of coefficients, the a_k, through (A5.2). The Fourier series for cosines and sines are the most common Fourier series, and they form orthogonal functions if integrated from 0 through π. These series are covered in sections A5.3 through A5.6, and they form the basis for computing laboratory L9. Confusion of the series and discrete Fourier expansions is likely in practical work because the integral is often approximated by a sum, as discussed in A4.4.

In *Fourier integral transforms*, also just called Fourier transforms, all the summations are replaced by integrals. Fourier integral transforms are discussed in A5.7, which provides the background for computing laboratory L10, wherein transformation from frequency to time domains is performed.

The order of presentation given here is not inevitable, and other authors (for example, Brigham, Wolf, and Wylie) provide alternative orderings. In what follows we choose, to be conventional and for convenience, the functions $\phi_k(x)$ to be the functions cosine, sine, or complex exponential, with parameters and ranges chosen to be appropriate for orthogonality. The weight factors ω_j will be chosen as unity, in order to obtain results with general applicability. Most of the formal results, even if not their interpretation, hold for much more general orthogonal functions. With the choice of the harmonic functions, we will be able to handle rapidly changing (even discontinuous) functions, as well as periodic functions.

A5.2 {1}   Discrete Fourier transforms

The first kind of Fourier expansion that we consider is the direct use of (A4.63) and (A4.67) with the trigonometric or complex-exponential functions. The algebra is much more straightforward for the latter, so we present it here. Furthermore, we assume that the x values are available at equally spaced points $x_j = jh$, with the origin of the x coordinates chosen so that $j = 1, 2, \ldots, N$.

Derivation of the discrete transform

Suppose that we try a Fourier expansion of the form

$$y(x) = \Sigma\, d_k \phi_k(x), \tag{A5.4}$$

where $\phi_k(x) = \exp(ikax)/\sqrt{N}$. The sum is over the index k, but its range is not yet specified. Neither has the quantity a yet been selected. The divisor $\sqrt{N}$ has been chosen so that derived formulas will be symmetric between the conversion from x to k and the inverse transformation from k to x. Can a and the range of summation be chosen so as to achieve orthogonality?

Consider the orthogonality sum S, corresponding to (A5.3), given by

$$S = \Sigma \exp(ikax_j) \exp(ilax_j)/N, \tag{A5.5}$$

with the sum over j from 1 to N. To evaluate S let $r = \exp(ipah)$, where $p = k + l$. Therefore, $S = \Sigma r^j/N = r(1 - r^N)/[(1 - r)N]$, where the last step comes from the geometric series formula in (A3.5). To get $S = 0$, we therefore need $r^N = \exp(iNpah) = 1$, that is $Nah = 2\pi$ for each p different from zero. Thus $a = 2\pi/(Nh)$ is the smallest value that makes the exponential functions orthogonal. For $p = 0$ ($l = -k$), $r = 1$ so that $S = \Sigma 1^j/N = 1$, since the sum ranges over N values of j.

Assuming that the weight factors ω_j are all unity, only the term in (A4.64) that has $l = -k$ will survive in the sum on the left side. Thus we have

$$a_{-k} = \Sigma y_j \phi_k(x_j). \tag{A5.6}$$

Therefore, changing k to $-k$, substituting the exponential form of ϕ_{-k}, using the formula for a, and renaming a_k by d_k for conformity with (A5.4), we have

$$d_k = \Sigma y_j \exp(-2\pi ijk/N)/\sqrt{N}, \tag{A5.7}$$

for the coefficients of the discrete Fourier transform in exponential form. In expression (A5.7) the summation ranges over the data index j from 1 to N. The range of k values is $-N$ to N.

► *Exercise* A5.2.1   Verify all the steps in deriving (A5.7) for the discrete Fourier transform, starting with the derivation of the orthogonality sum in (A5.5). □

The complementary expansion of the function in terms of the Fourier coefficients is, using (A5.4),

$$y(x) = \Sigma d_k \exp(2\pi ikx/Nh)/\sqrt{N}, \tag{A5.8}$$

in which the sum is now over k from $-N$ to N. Note that the Fourier coefficients d_k are complex numbers, even if the function y is purely real.

Properties of the discrete transform

Here we derive some of the properties of the discrete Fourier transform. Since many of these properties extend to the other Fourier expansions, such as Fourier series and Fourier integrals, it is a good idea to understand them at this point.

► *Exercise* A5.2.2   (The Wiener-Khinchin theorem)
Use equations (A5.7) and (A5.8) to prove the Wiener-Khinchin theorem, which

relates the Fourier coefficients to the original function: $\Sigma |d_k|^2 = \Sigma |y_j|^2$. On the left side the sum is over k from $-N$ to N, while on the right side the sum is over the data index j from 1 to N. $\square$

In the case that the k values refer to frequencies, and the x_j are therefore times, the two sides of the Wiener-Khinchin equality refer to the power frequency spectrum (on the left) and the time-averaged power (on the right).

Although the exponential form of the discrete Fourier transform is very convenient, highly symmetric, and even numerically convenient (as the fast Fourier transform algorithm and programs in A5.8 and L9.2 show), it is sometimes useful to have the corresponding trigonometric form of the series.

▶ *Exercise* A5.2.3   (Trigonometric form of discrete Fourier transform)
(a) To change equation (A5.8) from exponential to trigonometric form, rearrange terms so that the sum has $k \geqslant 0$, then combine the exponentials to show that

$$y(x) = \Sigma [a_k \cos (2\pi kx/Nh) + b_k \sin (2\pi kx/Nh)], \qquad (A5.9)$$

summed over $k = 0$ to N. Show that the coefficients of the cosine terms are given by

$$a_k = 2\Sigma [y_j \cos (2\pi kj/N)]/N, \quad k > 0, \qquad (A5.10)$$

but that

$$a_0 = \Sigma y_j/N, \qquad (A5.11)$$

and in both sums the range is over the data points from $j = 1$ to $j = N$. Show that the coefficients of the sine terms are

$$b_k = 2\Sigma [y_j \sin (2\pi kj/N)]/N. \qquad (A5.12)$$

(b) Show that the coefficient b_0 is irrelevant, because it is always multiplied by zero within the sum (A5.9). $\square$

Equations (A5.9) through (A5.12) provide the trigonometric form of the discrete Fourier transform. Note that, by convention, the factor N has been absorbed completely into the coefficients, rather than factored symmetrically as a square root. For those musically inclined, equation (A5.9) may be interpreted as stating that the tone is equal to the hum of its parts.

Note that there are two major restrictions to the discrete Fourier transform: (1) The values $y_j = y(x_j)$ must be obtained at equally spaced points x_j. This can be met in most analyses, especially if the y_j are experimental data and x_j is the independent variable in the experiment. In mathematical analysis the discrete Fourier transform is sometimes not a convenient Fourier expansion because the sums in computing the d_k, a_k, b_k cannot usually be performed analytically.

(2) The maximum frequency k for which coefficients are obtained is related to the number of data points N by $k \leqslant N$. This result is called *Nyquist's criterion*.[1] This condition on the maximum frequency can also be given by stating that at the maximum frequency sampled one cycle must include at least one data point, otherwise the oscillatory functions are not well enough defined. Note that the precise statement of the Nyquist criterion depends on whether the data are complex or real and on the range of k, particularly on whether it ranges over negative and positive values or only over positive values. See, for example, Otnes and Enochson, Chapter 1.8. In his book, *Science and Information Theory*, Brillouin discusses the relation between information content, frequency spectrum, and the Nyquist criterion. He also relates these topics to the thermodynamic concept of entropy that we discuss in L6.3.

The discrete Fourier transform is very suitable for data analysis, especially by computer, because the necessary sums can be easily programmed in loop structures. For example, in computing laboratory L9 electroencephalogram (EEG) data are analyzed by the discrete Fourier transform to convert a brain wave as a function of time to a brain wave as a function of frequency. Furthermore, recurrence relations can be used to simplify evaluation of the Fourier coefficients, as in the fast Fourier transform algorithm derived in A5.8. However, for working with analytic functions the Fourier series, considered next, is usually more convenient.

A5.3 {1}   Fourier series: Harmonic approximations

It is often convenient to obtain the Fourier coefficients by a procedure of integration, rather than summation, because integrals are often easier to evaluate analytically than are sums. This is accomplished by the transition from the discrete Fourier transform to the Fourier series.

From discrete transforms to series

Suppose that in the discrete Fourier transform we let the number of x values, N, become indefinitely large in such a way that $ax_j = 2\pi(j - 1)/N \rightarrow x$ and $a(x_{j+1} - x_j) = 2\pi/N \rightarrow dx$. Then the range of x is from 0 to 2π. The Fourier expansion (A5.8) becomes

$$y(x) = \Sigma \, d_k \exp (ikx)/\sqrt{(2\pi)}, \tag{A5.13}$$

in which the sum over k is from $-N$ to N. The Fourier series coefficients are obtained by applying the same limiting process to equation (A5.7). Thus

[1]Named after Harry Nyquist, a famous scientist in electrical engineering.

$$d_k = \int dx \, y(x) \exp(-ikx)/\sqrt{(2\pi)}, \qquad (A5.14)$$

for $k = -N$ to N. The range of integration is 0 to 2π.

▶ *Exercise* A5.3.1   Work through the steps from the discrete Fourier transform to the Fourier series in detail. □

The trigonometric form of the Fourier series is obtained by expanding the exponential in the preceding formulas into sines and cosines. You will find directly that

$$y(x) = \Sigma \, [a_k \cos(kx) + b_k \sin(kx)], \qquad (A5.15)$$

with the k sum running from 0 to N. In (A5.15) the Fourier coefficients of the cosine terms are

$$a_k = \int dx \, y(x) \cos(kx)/\pi, \qquad (A5.16)$$

for $k > 0$, and

$$a_0 = \int dx \, y(x)/2\pi. \qquad (A5.17)$$

The Fourier coefficients of the cosine terms are

$$b_k = \int dx \, y(x) \sin(kx)/\pi. \qquad (A5.18)$$

In equations (A5.16) through (A5.18) the range of integration over x is from 0 to 2π.

In formulas (A5.15) through (A5.18) we have bowed to convention and have not split the normalizing factors (here $\sqrt{\pi}$) between the function and its coefficients.

▶ *Exercise* A5.3.2   (Trigonometric series from exponential series)
Just to show how straightforward it is, derive (A5.15) through (A5.18) from the corresponding complex-exponential form of the Fourier series. □

The formulas (A5.15) through (A5.18) are often called *Fourier's theorem*.

Interpreting Fourier coefficients

Consider equations (A5.16) through (A5.18) for the Fourier coefficients. The integrals for a_k and b_k indicate to what extent $y(x)$ contains the oscillatory functions $\cos(kx)$ and $\sin(kx)$, respectively. These functions have a period of oscillation $2\pi/k$. Thus, as k increases, the more rapidly varying parts of $y(x)$ are the most significant in determining the corresponding a_k and b_k. In musical notation, $k = 1$ gives the *fundamental frequency*, and all the other frequencies are its *harmonics*.

▶ *Exercise* A5.3.3   Suppose that the Fourier series expansion (A5.15) has been used to compute the Fourier coefficients over the range of x values 0 to 2π. Show that we predict $y(x + 2\pi) = y(x)$. Thus, the Fourier series expansion generates a function that has the same period as the range of x values used to generate the coefficients. □

In our examples of Fourier series we consider periodic functions with rapid variation near some values of x and much slower variation near others. A delicate superposition of high and low harmonics is required to achieve this.

A5.4 {1} Examples of Fourier series

In this section we examine several examples of Fourier series that you are likely to use in applied science. The examples include square pulses, windows, and wedge functions.

Suppose that we want to make a periodic pulse sequence starting with the pure harmonic waves provided by an AC voltage supply. We consider pulses of various common shapes.

Square pulses

A frequently used pulse form is the square pulse shown in Figure A5.4.1. This pulse has unit amplitude and period 2π, and it is antisymmetric about $x = \pi$. The amplitude restriction can always be removed by applying a multiplying scale factor if the amplitude differs from 1. In A5.6 you will learn how to determine Fourier coefficients for arbitrary intervals.

The square pulse is intractable with the Taylor series from A3.3 (based on differentiation), because of the jump at π. However, with Fourier series (based on integration), such a jump is no obstacle. The function to be least-squares fitted by cosines and sines is described by

$$y(x) = \quad 1, \quad 0 < x < \pi, \tag{A5.19}$$

$$y(x) = -1, \quad \pi < x < 2\pi. \tag{A5.20}$$

Using (A5.14) we obtain directly for the Fourier series coefficients

$$d_k = \int dx \, y(x) \exp(-ikx)/\sqrt{(2\pi)}, \tag{A5.21}$$

that is

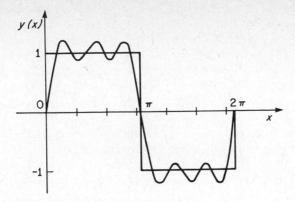

Figure A5.4.1   Square pulse, shown for the range 0 to 2π, and its Fourier series expansion (A5.28) using the first three non-zero terms.

$$d_k = \int dx \exp{(-ikx)}/\sqrt{(2\pi)} - \int dx \exp{(-ikx)}/\sqrt{(2\pi)}. \quad \text{(A5.22)}$$

In the first form of this integral the limits on x are 0 to 2π. In the second form the value of $y(x)$ in each interval has been taken into account, so that the limits of integration are 0 to π and π to 2π, respectively. Working out in detail both of the integrals in (A5.22), you will find the Fourier series coefficients for the square pulse

$$d_k = \sqrt{(2/\pi)}[1 - (-1)^k]/ik. \quad \text{(A5.23)}$$

If we change the index from k to K in order to sum over only nonzero terms, this can be written compactly as

$$d_{2K} = 0, \; d_{2K+1} = 2\sqrt{(2/\pi)}/[i(2K + 1)]). \quad \text{(A5.24)}$$

▶ *Exercise* A5.4.1   Verify the steps (A5.21) through (A5.24) in deriving the Fourier coefficients for the square pulse by working out each integral in detail. ☐

The Fourier series expansion of the square pulse is thus

$$y(x) = 2\Sigma \exp{[i(2K + 1)x]}/[(2K + 1)\pi i], \quad \text{(A5.25)}$$

in which the summation is over $K = -[N/2]$ to $K = [N/2]$, where $[t]$ denotes the truncated integer value of t.

As it stands, formula (A5.25) is not very convenient for numerical work, because $y(x)$ is real-valued, but complex exponentials are apparent on the right-hand side. It really takes only a little exercise to change all this.

▶ *Exercise* A5.4.2   Use the complex-exponential form of the sine function, from A2.3, to show that for the square pulse of unit amplitude and period 2π equation (A5.25) can be reduced to

$$y(x) = 4\Sigma \sin[(2K + 1)x]/[\pi(2K + 1)], \qquad (A5.27)$$

in which the summation is over K from zero to $[N/2]$. ☐

Explicitly, the Fourier series approximation for the first M nonzero terms of the square pulse of unit amplitude and period 2π is

$$y(x) \approx 4\{\sin(x) + \sin(3x)/3 + \ldots + \sin[2M - 1)x]/(2M - 1)\}/\pi \qquad (A5.28)$$

Fourier coefficients and symmetry conditions

To explain why most of the Fourier coefficients for the square pulse vanish, we can use symmetry conditions. The function considered satisfies $y(x + \pi) = -y(x)$, so that it can be formed only out of functions that have the same property under $x \rightarrow x + \pi$. This restricts the expansion to the sine functions.

In general, unnecessary effort may often be avoided by using symmetry conditions *before* the integrals for the coefficients are attempted. At least one should verify that the final result is consistent with the symmetry conditions. Problem [A5.2] requires use of symmetry conditions.

▶ *Exercise* A5.4.3   (Numerical Fourier series for square pulse)
(*a*) Write a program, for computer or pocket calculator, that inputs the integer M, then computes the Fourier series expansion for the square pulse shown in Figure A5.4.1, using the first M nonzero Fourier series coefficients. Suitable values might be $M = 2, 3, 6$.
(*b*) Plot the exact square function on a high-resolution plotter (such as hand drawing on accurate graph paper), and on this superimpose the prediction from the Fourier series with the M terms. For $M = 3$ it should resemble Figure A5.4.1. How quickly does the series approximation approach the square as M increases?
(*c*) If your plotter is of low resolution, also plot the *difference* (including sign) between the exact function and its Fourier series approximation. For example, the printer-plot program provided in computing laboratory L4 will not have enough resolution to give a good representation of the Fourier expansion past about 6 terms, but the difference (displayed on an appropriate scale) will be a more sensitive indicator. ☐

The wedge function

The next function that we consider for Fourier series analysis is the wedge, or isosceles triangle, function. The Fourier series expansion of the wedge func-

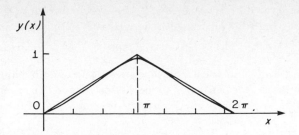

Figure A5.4.2   Wedge function for a Fourier series expansion. The curve is the Fourier series approximation (A5.32) for three non-zero terms.

tion shown in Figure A5.4.2 is of interest, especially for its application in image enhancement. (See, for example, the text by Pratt.) The wedge function is defined by

$$y(x) = x/\pi, \ 0 \leqslant x \leqslant \pi, \tag{A5.29}$$

$$y(x) = (2\pi - x)/\pi, \ \pi \leqslant x \leqslant 2\pi. \tag{A5.30}$$

The complex-exponential form of the Fourier series for the wedge function is readily calculated by dividing the integral into two regions, similarly to the square pulse. We find for the complex Fourier series expansion of the wedge function

$$y(x) \approx 1/2 - (2/\pi^2)\Sigma \exp \left[i(2K + 1)x\right]/(2K + 1)^2, \tag{A5.31}$$

in which the summation is over K from zero to the number of terms desired in the approximation.

► *Exercise* A5.4.4
(*a*) Derive the Fourier series expansion (A5.31) for the wedge shape in detail.
(*b*) Convert (A5.31) to trigonometric form, to show that the trigonometric Fourier series expansion of the wedge function is

$$y(x) \approx \tfrac{1}{2} - (4/\pi^2)\Sigma \cos \left[(2K + 1)x\right]/(2K + 1)^2. \tag{A5.32}$$

Here K runs from 0 to $M - 1$, where M is the number of non-zero terms desired for the approximation. Figure A5.4.2 shows the expansion for $M = 3$, which has a good fit because of the rapid convergence of the series (A5.32).
(*c*) Use symmetry conditions to explain why only cosine terms of odd harmonics can appear in the wedge expansion. □

Now that we have a Fourier series for the wedge function, given by (A5.32), it will be interesting for you to see how well this expansion matches the original function.

▶ *Exercise* A5.4.5   (Numerical Fourier series for wedge function)
(*a*) Write a program for computer or pocket calculator that inputs the integer M, then uses the first M terms in the Fourier series expansion of the wedge function in Figure A5.4.2.
(*b*) Graph the wedge function and superimpose the Fourier series approximation for each of your choices of M. How quickly does the series approximation approach the wedge as M increases? Note that if your plotter is of low resolution, then it is also worthwhile to plot the difference between the exact function and its Fourier series approximation. Do your results resemble the fit given in Figure A5.4.2 for $M = 3$?
□

Is there a relation between the Fourier expansion of the square pulse and that of the wedge? One connection between the two shapes is that the square pulse is the derivative of the wedge. Does the connection hold term by term in their Fourier series? To find out, please try Problem [A5.1].

Window functions

A window, which allows a signal in a certain range of x, but blocks out the signal outside this range, is a very important function in signal processing. We consider a very simple window; more general classes of window functions are covered in the books by Oppenheim and Schafer, by Papoulis, and by Pratt. The window function is also called the "boxcar" function, from its resemblance to the silhouette of a railroad boxcar.

Our third example of a Fourier series will shed more light on such series. The window function of unit height is shown in Figure A5.4.3. We could again perform the Fourier integrals. But, knowing the square-pulse series, we can readily derive the series for the window function by noting that if we add unity to the square pulse, then halve the result, we have the window function, except that it is $\pi/2$ ahead of the window that we have drawn. Therefore, the Fourier series for the window is obtained from that for the square pulse, (A5.25), by evaluating it at $x - \pi/2$, dividing it by 2, then adding it to $\frac{1}{2}$. That is, for the window function the complex exponential form of the Fourier series expansion is

$$y(x) \approx \tfrac{1}{2} + (1/\pi i)\Sigma \exp\left[i(2K + 1)(x - \pi/2)\right]/(2K + 1), \quad \text{(A5.33)}$$

while the trigonometric form of the window function is

$$y(x) \approx \tfrac{1}{2} - (2/\pi)\Sigma(-1)^K \cos\left[(2K + 1)x\right]/(2K + 1). \quad \text{(A5.34)}$$

In both these expansions the summations over K run from 0 to $M - 1$.

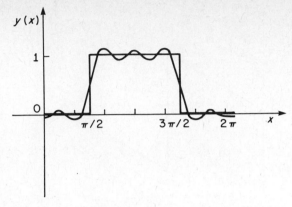

Figure A5.4.3   Window function with its Fourier series expansion for the first three non-zero terms in (A5.34).

The window function is so-called because when it is used to multiply a function it selects from that function only that part lying between $\pi/2$ and $3\pi/2$. The extension to windows of other dimensions and centers is made through the transformation explained in A5.6.

One disadvantage of the window function is that it makes a very sudden cutoff, which introduces undesirable oscillations into the Fourier series expansion of the windowed function. You can investigate this problem (which is similar to the Gibbs phenomenon discussed in A5.5) in more detail in Problem [A5.3].

Convergence of Fourier series

A question of considerable practical importance in generating various pulse forms, such as those just considered, by superposition of harmonics in a Fourier series is "How many harmonics are required to give an accurate representation of the pulse form?" We study this for the example of the sawtooth function shown in Figure A5.4.4. We first require the Fourier series for the sawtooth; try extracting it as an exercise.

▶ *Exercise* A5.4.6   Show that the Fourier expansion for the sawtooth function in Figure A5.4.4 is

$$y(x) \approx (2/\pi)\Sigma \, (-1)^{k-1} \sin (kx)/k, \qquad (A5.35)$$

in which the sum over k is from 1 to M, the number of terms in the sum. □

For a given value of M, how linear is the fit produced by the Fourier series expansion (A5.35)? We know from the general treatment in A4.7 and in A5.1

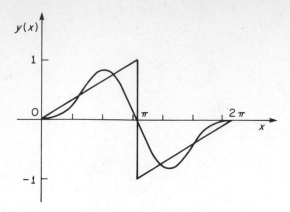

Figure A5.4.4.   Sawtooth function and the approximation to it obtained with the first two terms of the Fourier
series (A5.35).

that the Fourier expansion gives the best fit in the least-squares sense. However, this does not guarantee that the fit is particularly good in any small
region of x values. You may investigate this in the following exercise.

▶ *Exercise* A5.4.7   (Properties of the sawtooth function)
Consider the portion of the sawtooth shown in Figure A5.4.4 in the range
$0 \leqslant x \leqslant \pi$. In this range $y(x) = x/\pi$, since we have chosen the scale of y so that it is
unity at $x = \pi$.
(*a*) Show that the Fourier series predicts $y(\pi) = 0$. Is this consistent with the sawtooth?
(*b*) Show that the Fourier series predicts

$$y(\pi/2) = (2/\pi)\Sigma \, (-1)^K/(2K + 1), \qquad (A5.36)$$

in which the summation is from $K = 0$ to $K = [M/2]$, the truncated integer part of
$M/2$.
(*c*) Use equation (A5.36) to show that $M = 6$ is required to produce 5% accuracy in
the prediction that $y(\pi/2) = 0.5$.
(*d*) (For those interested in numbers and insight.) Use equation (A5.36) to derive a
series expansion for π, then write a simple program for pocket calculator or computer to estimate π by summing the Fourier series to some large value of M. Do you
think that this series provides a numerically reliable method for estimating π? Consider your answer in terms of the discussion of numerical noise and subtractive cancellation in A4.2. □

This example on the convergence of Fourier series is very important in electronic circuit design for sawtooth, or "ramp," voltages. Such voltages are usually generated from the AC mains frequency and harmonics of it. The fewer

the harmonics that have to be combined to give a suitably linear ramp, the simpler the circuit design.

A5.5 {2}   Diversion: The Gibbs phenomenon

If we look at the Fourier series approximation to a function with a discontinuity, as in the examples in A5.4, we see that there is an overshoot of the Fourier series near the discontinuity. This phenomenon was first explained by J. Willard Gibbs, one of the founders of theoretical physics in America.[1] Gibbs, in correcting an error in one of his earlier published remarks on Fourier series,[2] pointed out the origin of this overshoot, now known as the Gibbs phenomenon. The history of the effect is also discussed by Bennett in Chapter 7 of his book.

We examine the Gibbs phenomenon as follows, using the example of the square pulse shown in Figure A5.4.1. The discrepancy between the Fourier series to M non-zero terms and the square pulse is obtained from (A5.28) as

$$\Delta_M(x) = -4\Sigma \sin\left[(2K + 1)x\right]/\left[(2K + 1)\pi\right], \qquad (A5.37)$$

in which the sum is over K values greater than M, since these terms must be included if the square pulse shape is to be reproduced. The summation can be approximated by an integral, so that

$$\Delta_M(x) \approx 2[-\int dt \, \sin(t)/t]/\pi, \qquad (A5.38)$$

in which the integral over t extends from $(2M + 1)x$ to ∞. The integral (including the negative sign), is the si function, discussed in Chapter 5 of Abramowitz and Stegun. It has its first maximum at $x = \pi/(2M + 1)$, and its value there is 0.281. Thus the fraction by which the Fourier series expansion to M terms overshoots the square pulse near $x = \pi$ tends to a constant value $2*0.281/\pi = 0.18$ for large M, but the width of this overshoot tends to zero as $1/M$, so that its area tends to zero. Therefore, for large enough M, the Gibb's phenomenon does not have serious consequences, provided that one stays away from the discontinuities of the function.

Problem [A5.4] considers the more general question of damping out oscillatory error by means of Lanczos damping factors. In the terminology of signal processing, the damping is provided by a low-pass filter.

Why worry about this overshoot behavior? One reason is that any rapidly changing or discontinuous function will suffer similarly and can be similarly cured. The phenomenon exists partly because we are making a least-squares fit for the Fourier expansion (see A4.7 and A5.1), and the best one can do is to sacrifice near the discontinuity in order to achieve a good fit elsewhere.

[1]See R. J. Seger, *J. Willard Gibbs*, Pergamon Press, Oxford, 1974.
[2]J. W. Gibbs, *Nature*, **59**, 606 (1899).

A5.6 {2}   Fourier series for arbitrary intervals

Thus far in considering Fourier expansions we have assumed functions in the range $0 \leqslant x \leqslant 2\pi$, and of unit maximum magnitude. It is straightforward to include both arbitrary intervals and arbitrary maximum magnitudes.

Interval scaling

Suppose, as usual in Fourier analysis, that we have a function periodic in x for $x_- \leqslant x \leqslant x_+$. We consider the problem of relating the Fourier series for a function defined in this range to that for $0 < x < 2\pi$. To convert to the range 0 to 2π scale x as

$$x' = 2\pi(x - x_-)/(x_+ - x_-) \tag{A5.39}$$

which ranges over 0 to 2π as x ranges over x_- to x_+. Also note that

$$dx' = 2\pi dx/(x_+ - x_-). \tag{A5.40}$$

In terms of the x' variable we now have the Fourier series for an arbitrary interval

$$y(x') = \Sigma \, d'_k \exp(ikx') = \Sigma \, d'_k \exp[i2\pi k(x - x_-)/(x_+ - x_-)], \tag{A5.41}$$

in both of which the summation is over the same range of the index k, and where

$$d'_k = \int dx \, y(x) \exp[-2\pi ikx/(x_+ - x_-)]/(x_+ - x_-), \tag{A5.42}$$

in which the integral now ranges over x_- to x_+.

▶ *Exercise* A5.6.1   Use the Fourier series coefficient formula (A5.14) to derive expression (A5.42) for the Fourier coefficients over the range x_- to x_+ ☐

▶ *Exercise* A5.6.2   Convert the exponential form of the Fourier series (A5.42) to trigonometric form to show that for the range x_- to x_+ the Fourier coefficients are

$$a_0 = \int dx \, y(x)/(x_+ - x_-) \tag{A5.43}$$

$$a_k = 2\int dx \, y(x) \cos[2\pi k(x - x_-)/(x_+ - x_-)]/(x_+ - x_-), \quad k > 0, \tag{A5.44}$$

$$b_k = 2\int dx \, y(x) \sin[2\pi k(x - x_-)/(x_+ - x_-)]/(x_+ - x_-). \tag{A5.45}$$

In each of these integrals the range of x is x_- to x_+. ☐

As a final confirmation of the correctness of our interval scaling procedure for Fourier series, it is interesting to show that the Fourier series is periodic over the new interval.

▶ *Exercise* A5.6.3   Show that $y(x)$ determined by the Fourier expansion (A5.41) is periodic in x with period $(x_+ - x_-)$. ☐

Magnitude scaling

Scaling of the function $y(x)$ by multiplying it throughout by the constant scale factor s merely multiplies each Fourier coefficient by the same factor s, because the Fourier coefficients are linearly related to the function y. This is clear from the origins, in A4.7 and A5.1, of the Fourier expansion as a *linear* least-squares expansion in complex-exponential or circular functions. Therefore, in each of the examples of Fourier series in A5.4, the Fourier coefficients of any multiple of the square, wedge, window, or sawtooth functions are obtained by using the same multiplying factor.

A5.7 {3}   Fourier integral transforms

Suppose that we want to describe, for example, an impressed force or a voltage pattern that is not periodic as a function of time in terms of its frequency components. Can the Fourier series be adapted to this situation?

One way in which this can be done is to choose x_- and x_+ so that they lie far outside the range of x values for which we will use the Fourier series. The periodicity of the Fourier series will then be inconsequential. This idea also leads to the discrete harmonics k in the Fourier series becoming continuous frequency variables.

From Fourier series to Fourier integrals

Our choice in deriving the Fourier integral transform is to set $x_- = -L$ and $x_+ = L$, and to eventually let $L \to \infty$. By this means we produce the Fourier integral transform.

Most examples using Fourier integral transforms relate to time and temporal frequency. Therefore we change the independent variable from x to t for the rest of this section. The notation and results will then correspond to the majority of other presentations of Fourier transforms.

With the change of variable from x to t, $t_+ - t_- = 2L$, and we can make a change of variable to obtain the angular frequency of the kth term as

$$\omega_k = \pi k / L. \qquad (A5.46)$$

The difference between successive frequencies is thus

$$\Delta\omega = \pi/L. \tag{A5.47}$$

The Fourier series (A5.41) in this new notation can then be written as

$$y(t) = \Sigma\, Y(\omega_k)\, \exp\, (i\omega_k t)\Delta\omega/\sqrt{(2\pi)}, \tag{A5.48}$$

in which the sum over k is from $-N$ to N. The Fourier coefficients in expression (A5.48) are given by, on using (A5.42),

$$Y(\omega_k) = \int dt\, y(t)\, \exp\, (-\omega_k t)/\sqrt{(2\pi)}, \tag{A5.49}$$

in which the integral ranges over $-L$ to L. In these two expressions the factor $\sqrt{(2\pi)}$ has been split between the two formulas to achieve symmetry.

As we let $L \to \infty$, the spacing $\Delta\omega \to 0$ and the Fourier sum (A5.48) approaches an integral.[1] The label k can also be dropped, because the corresponding variable ω becomes continuous rather than discrete. Thus we obtain the Fourier integral transform pair

$$y(t) = \int d\omega\, Y(\omega)\, \exp\, (i\omega t)/\sqrt{(2\pi)}, \tag{A5.50}$$

$$Y(\omega) = \int dt\, y(t)\, \exp\, (-i\omega t)/\sqrt{(2\pi)}, \tag{A5.51}$$

in which both integrals range from $-\infty$ to ∞.

This pair of equations is to be used in a complementary manner; if the frequency response $Y(\omega)$ is known, then the time response $y(t)$ can be computed using the first equation, and vice versa using the second equation. By reverting to the x notation for spatial coordinates, we see that wave number response $Y(k)$ and spatial response $y(x)$ are similarly related. (Wavenumber k is related to wavelength λ by $k = 2\pi/\lambda$.)

In general, in order to form a Fourier transform pair the variables such as ω and t, or k and x, must be dimensionally reciprocal so that the exponents in the Fourier transform relations are dimensionless.

▶ *Exercise* A5.7.1   Suppose that you are given the expression

$$y(x,t) = \int dk \int d\omega\, Y(k,\omega)\, \exp\, [i(kx - \omega t)], \tag{A5.52}$$

what is a suitable physical interpretation? For background on this question see French's book on vibrations and waves. □

---

[1]For a mathematically rigorous derivation of this result, and the necessary conditions, see Churchill.

The symmetric form of the Fourier integral transform pair (A5.50) and (A5.51) is very similar to that of the discrete Fourier transform pair (A5.8) and (A5.7) in section A5.2. Indeed, by applying the limiting process to both j and k in the discrete-transform relations, we could derive the Fourier integral transform without the intermediary of the Fourier series. Conversely, the integral transforms could be discretized to produce the series results. The variety of approaches in various texts can be very confusing, both to students and to authors!

Applying Fourier integral transforms

The Fourier integral transform is most useful when applied to analytical problems for which the pair of integrals can be evaluated in closed form. For example, the solution of differential equations may be greatly simplified by using Fourier integral transforms, as shown in A7.4.

In data analysis the Fourier integral transform is usually approximated or replaced by the discrete Fourier transform presented in A5.2, because the data are obtained at discrete (often equal) spacing. For example, the discrete Fourier transform is used in the analysis of EEG patterns in computing laboratory L9, while the Fourier integral transform is applied to the analysis of the ammonia inversion resonance line width in L10.

As an analytical example of the Fourier integral transform, consider the triangle-shaped function symmetric about the origin, as shown in Figure A5.7.1:

$$y(t) = (T + t)/T^2, \quad -T \leqslant t \leqslant 0, \tag{A5.53}$$

and

$$y(t) = (T - t)/T^2, \quad 0 \leqslant t \leqslant T. \tag{A5.54}$$

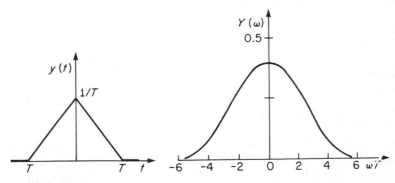

Figure A5.7.1 Triangle function (left), equations (A5.53) and (A5.54), and its Fourier integral transform (right) given by (A5.55).

Note that $y(t)$ is scaled so that the area of the triangle is unity for any value of T. If we substitute the expression for $y(t)$ into (A5.49), separate the integral into two integrals, one for negative t and the other for positive t, then evaluate the two integrals directly, we find for $\omega \neq 0$ the Fourier integral transform of the triangle function

$$Y(\omega) = [\sin (\omega T/2)/(\omega T/2)]^2/\sqrt{(2\pi)}. \tag{A5.55}$$

▶ *Exercise* A5.7.2   Show the integration steps leading to this result for the Fourier integral transform of a triangle shape. (This is straightforward but tedious.) ☐

A difficulty arises if we try to substitute $\omega = 0$ directly in this result, because an indeterminate form, 0/0, results. The next exercise suggests two ways out of this problem.

▶ *Exercise* A5.7.3
(a) Show directly from the integral definition that for $\omega = 0$ the Fourier integral transform of any function $y(t)$ is given by

$$Y(0) = [\text{area under } y(t)]/\sqrt{(2\pi)}, \tag{A5.56}$$

and therefore that for the triangle function $Y(0) = 1/\sqrt{(2\pi)}$.
(b) Demonstrate, using a Maclaurin expansion (see A3.4) of the sine function in (A5.55), that $Y(0)$ is correctly obtained by letting ω tend to zero in this equation. ☐

The next exercise will give you some practice in Fourier transform numerical properties.

▶ *Exercise* A5.7.4   Calculate then sketch the Fourier integral transform (A5.55) of the triangle function in Figure A5.7.1 and verify that:
(1) $Y(\omega)$ is symmetric about $\omega = 0$.
(2) $Y(\omega)$ falls to half its maximum value at $\omega = 2.78/T$.
(3) The zeros of $Y(\omega)$ occur at $\omega = 2n\pi/T$, n a nonzero integer.
(4) A secondary maximum of $Y(\omega)$ occurs near $\omega = 3\pi/T$, and its height is about 0.045 that of the first maximum. ☐

The results from exercise (A5.7.4) are important in the theory of wave diffraction from apertures. See, for example, French's book on vibrations and waves or Pippard's book on the physics of vibration.

You can get more experience with Fourier integral transforms by trying Problem [A5.5] on the Dirac delta function and by working computing laboratory L10 on the analysis of resonance line widths. Section L10.3 presents the Fourier integral transform of the Lorentzian function, which often gives a good representation of the frequency dependence of resonance response.

A5.8 {3}   A fast Fourier transform algorithm

We have thus far considered formulas for the various Fourier expansions, but we have not described any practical algorithms for their evaluation. Discrete Fourier transforms similar to those considered in A5.2 are now very widely used in data analysis and digital signal processing because a very efficient algorithm, the fast Fourier transform (FFT), has been developed for their evaluation. The nomenclature of the FFT is misleading; the algorithm applies to the discrete Fourier transforms in A5.2, *not* to the Fourier integral transform in A5.7.

Derivation of FFT algorithm

We present here a straightforward, complete, and, it is hoped, clear derivation of the FFT algorithm. Our starting point is similar to the discrete Fourier transform formulas in A5.2. We assume an expansion of the form

$$y(x) = \Sigma\, c_k \exp\,(2\pi ikx/Nh)/\sqrt{N}, \tag{A5.57}$$

in which the sum is over k from 1 to N. The data are assumed to be available at equally spaced intervals h, beginning at x_1. That is $x_j - x_1 = (j - 1)h$.

This expansion differs from that in A5.2 because the sum is only over positive k values, rather than over k from $-N$ to N. This can readily be accommodated, as the following exercise shows.

▶ *Exercise* A5.8.1
(*a*) Show that if the average value (DC level) of the data values y_j is forced to be zero (for example, by subtracting the average value from each of the original data), then $d_0 = 0$ in the Fourier analysis presented in A5.2.
(*b*) Use the complex conjugate of (A5.8), assume that y is real, then rearrange the summation index k, to show that for real data

$$d_k = (c_k + c_{-k}{}^*)/2. \tag{A5.58}$$

(*c*) Repeat the orthogonality arguments given in A5.2 to show that

$$c_k = \Sigma\, y_j \exp\,(-2\pi ijk/N)/\sqrt{N}, \tag{A5.59}$$

in which the summation over j is from 1 to N. □

To continue our derivation of the FFT algorithm applied to the expansion (A5.57), we can write briefly

$$C_k = \Sigma\, y_j E(N)^{jk}, \tag{A5.60}$$

where $C_k = c_k/\sqrt{N}$, and

$$E(N) = \exp(-2\pi i/N). \tag{A5.61}$$

The number of arithmetic operations required to compute all the Fourier coefficients from $k = 1$ to $k = N$ directly from this equation increases at least as fast as N^2 because each coefficient requires combining N values of the y_j according to (A5.59), and there are N coefficients. We call such a direct evaluation of the Fourier coefficients a *conventional Fourier transform* (CFT) algorithm.

Suppose that the calculation of the set of coefficients for the discrete Fourier transform (A5.57) could be divided into two similar subcalculations each of length $N/2$. The calculation would then be reduced to roughly $2(N/2)^2$, which is a factor of 2 smaller than N^2.

Assuming that N is even, one way of dividing the Fourier coefficient calculation problem is to combine the odd values $j = 1,\ldots,N - 1$ (these are of the form $2r - 1$ with $r = 1,\ldots,N/2$), then the even values $j = 2,\ldots,N - 2,N$ (of the form $2r$ with $r = 1,\ldots,N/2$), as shown by the two circular arcs for $i = 2$ in Figure A5.8.1. The formula (A5.60) for C_k can be written compactly as

$$C_k = \Sigma E(N)^{(m-1)k} \Sigma y_{2r+m-1} E(N/2)^{rk}. \tag{A5.62}$$

The first summation is for $m = 0$ and $m = 1$. The inner summation is over r from 1 to $N/2$, a factor of 2 shorter than the original sum in (A5.57). In this equation we have used the distributive property of the complex exponential

$$E(N)^{a+b} = E(N)^a E(N)^b, \tag{A5.63}$$

and

$$E(N)^{pa} = E(N/p)^a \tag{A5.64}$$

for $p \neq 0$. Note also that

$$E(N)^0 = E(N)^N = 1, \quad E(N)^{-1} = \exp(2\pi i/N). \tag{A5.65}$$

▶ *Exercise* A5.8.2   Prove in detail the three preceding properties of the complex exponential $E(N)$ that are needed in developing the FFT algorithm. See also Problem [A2.3]. □

By this level of division ($i = 2$ in Figure A5.8.1) there are now two discrete Fourier transforms each of half the previous length. This process of halving each transform can be continued as long as the number of points to be divided in each group at each level is divisible by 2. For example, in Figure A5.8.1 at level $i = 3$ there are 4 groups each with only 2 points in each. Note that

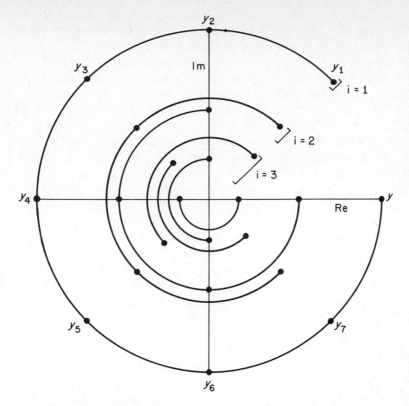

Figure A5.8.1   Division of the FFT into a sequence of smaller problems, shown in the complex plane for sample size 8. The location of each point in the complex plane is appropriate for the fundamental frequency of real data.

within each group along an arc the relative phase between points is the same as at the outer level $i = 1$.

In particular, we restrict ourselves to cases in which the total number of data points N is a power of 2,

$$N = 2^{\nu}, \tag{A5.66}$$

where ν is integral. Then the discrete Fourier transform can be halved ν times, ending with $N/2$ subintervals each of length 2.

Reordering the FFT coefficients

At each step of the fast Fourier transform the values of the C_k become reordered because of the division into odd and even terms. This reordering is also called *bit-reversal*. To understand this reordering, consider the example in Figure A5.8.1 for $N = 8$. At level $i = 3$, moving inwards on the semicircular arcs, the subscripts of the y values are 1,5,3,7,2,6,4,8.

When we sort a sequence into odd and even terms, the odd terms all have the same least-significant bit in a binary representation, and all the even terms have the same (other) least-significant bit. For example, for y_4, $4 - 1 = 3 = 2^2 0 + 2^1 1 + 2^0 1$, which in binary representation is 011. The sorting that occurs from level $i = 1$ to $i = 2$ sorts all those y_j with $j - 1$ having least-significant bit zero into the outer arc for $i = 2$, and all those with $j - 1$ having least-significant bit unity are sorted into the inner arc for $i = 2$. Similarly, from level $i = 2$ to $i = 3$ a subsequencing occurs. The final result after v levels is that the binary representation of $k - 1$ for each C_k is in bit-reversed order, as one sees in Figure A5.8.1 by comparing levels $i = 1$ and $i = 3$.

The final step in an FFT algorithm is therefore to reorder the Fourier coefficients according to this bit-reversed labeling scheme. The FFT programs in Pascal and in Fortran that are presented in L9.2 perform the bit reversal by a computer-independent method in procedure (or function) BITREV. By working through the next exercise you will learn how the bit-reversal algorithm works.

▶ *Exercise* A5.8.3   To understand how bit reversal in FFT is implemented in procedure BITREV in L9.2, consider the binary representation of an integer j between 0 and $2^v - 1$ (which is the range of the index $k - 1$ for the C_k). Such an integer can be represented by

$$j = j_0 + 2j_1 + \ldots + 2^{v-2}j_{v-2} + 2^{v-1}j_{v-1}, \tag{A5.67}$$

in which each j_i is either 0 or 1. The bit-reversed integer is then j_R, where

$$j_R = 2^{v-1}j_0 + \ldots + j_{v-1}. \tag{A5.68}$$

Show that j_R can be generated similarly to Horner's algorithm (see A4.2), as follows:

```
jT : = j; jR : = 0
    Iterate the next steps in integer arithmetic
    For i : = 1 to v
        jD : = jT/2
        ji-1 : = jT - 2*jD
        jR : = 2*jR + ji-1
        jT : = jD
```

□

Now that we have derived the FFT algorithm, was it worth all the effort?

Comparing FFT with conventional transforms

Here we compare the time for calculating a discrete Fourier transform of N points, using the fast (FFT) and conventional (CFT) algorithms.

For the FFT the total number of arithmetic steps of the form (A5.62) is roughly proportional to that required for N steps for each of v subintervals, that is to $Nv = N\lg_2 N$. A more-detailed analysis (Brigham, Chapter 12) shows that the time $T(FFT)$ for the fast Fourier transform of N points in our base-2 algorithm is about

$$T(\text{FFT}) = 5T_a N\lg_2 N, \tag{A5.69}$$

where T_a is the time for a typical real arithmetic operation, such as multiplication or addition.

For the CFT, the time T(CFT) to compute N coefficients by the direct application of (A5.59) is shown by Brigham to be about

$$T(CFT) = 6T_a N^2. \tag{A5.70}$$

The great increase in speed of the FFT over the CFT, especially for significant values of N, is made clear in the following exercises.

▶ *Exercise* A5.8.4   (Comparison of speeds of FFT and CFT)
(a) Using equations (A5.70) and (A5.69), make a graph of the relative times for CFT and FFT algorithms as a function of N. For example, use N = 8,16,512. An appropriate scale for N is logarithmic to base 2, that is $\lg_2 N$ = v.
(b) On my personal computer I programmed and ran the FFT algorithm program given in L9.2. I found that for $N = 8$, $T(\text{FFT}) = 3.5$ sec, and for $N = 16$, $T(\text{FFT}) = 7.5$ sec. From these data, estimate a range for T_a and thereby estimate a lower and an upper limit on the FFT time (in minutes) for a data sample of $N = 1024 = 2^{10}$ points.
(c) Use the same estimates of T_a to show that the CFT would take about 2 days of continuous running to Fourier transform the same 1024 data points. This is a factor of 120 times longer than the FFT. ☐

It is not impractical to use 1024 points in a discrete Fourier transform, as Problem [A5.6] shows. Computing laboratory L9.2 implements the FFT algorithms as computer programs in Pascal and in Fortran. You might use these, as I did, for a direct timing of the speed of the FFT in your computer system.

Further reading on the fast Fourier transform and its wide-ranging applications is provided by the monograph by Brigham, in Chapter 12 of Papoulis's book on signal processing, in Chapter 6 of Oppenheim and Schafer, and in Chapter 3 of Wolf.

Problems on Fourier expansions

[A5.1] {2}   (Relating Fourier coefficients through algebra) Fourier expansion coefficients of a given function can sometimes be obtained from those of another function by finding geometric or analytic relations between the functions. The same relations are then applied to the terms in the series expansion. As an example of this method applied to Fourier series, consider the relation between the wedge and the square pulse.
(a) Show that if you take the derivative with respect to x of the wedge function considered in (A5.29), (A5.30), then multiply it by π, you obtain the square pulse considered in (A5.19), (A5.20).
(b) To check the formulas obtained for these functions, differentiate the Fourier series (A5.31) for the wedge term by term and verify that multiplication of each coefficient by π regains the Fourier series for the square pulse, (A5.25).

[A5.2] {1}   (Relating Fourier coefficients through geometry) Relate the square pulse and the sawtooth function Fourier series by means of the following geometric correspondence.
(a) Show from the graphs of the two functions, Figures A5.4.1 and A5.4.4, that by reflecting the sawtooth $V_{s-}(t)$ about $t = 0$, then displacing it in t by π, we produce a function that, when added to the original sawtooth, produces the square-pulse form $V_{sp}(t)$.
(b) Use the Fourier series expansions for the sawtooth and square pulse, equations (A5.35) and (A5.27), to show that the construction in (a) is correct.

[A5.3] {2}   (Smoothing window functions) A window is a (usually) rectangular structure through which a signal (typically photons) can pass, but signals from beyond the window are blocked. From simple diffraction experiments, as discussed in the book by French, you may recall that if the signal wavelength is comparable to the window width, then the signal transmitted is blurred near the window edges. This effect is also present in the Fourier series expansion of a square pulse, which can also be viewed as a window, as discussed in A5.4.

To understand windowing in more detail, consider the smoothing function $h(x) = [1 - \cos(x)]/2$. Show that the Fourier series expansion of the product of this smoothing function with the window function shown in Figure A5.4.3 is much smoother than that of the original window function, especially if only a small number of terms is used in the series. For more on windowing functions and their applications in signal processing, read Oppenheim and Schafer, Papoulis, and Wolf.

[A5.4] {3}   (Gibbs phenomenon and Lanczos damping factors) The Gibbs phenomenon discussed in A5.5 can be alleviated by the following modification to the Fourier series expansion coefficients for a square pulse, or other sharp-edged function.

(*a*) Show that if the Fourier expansion of the square pulse to M terms is integrated over a range $\pm\pi/(2M)$ centered on x, then the kth Fourier coefficient of the resulting function is multiplied by the Lanczos damping factor

$$\sigma_k = \sin{(\pi k/2M)}/(\pi k/2M). \tag{A5.71}$$

(*b*) Calculate and graph the square-pulse Fourier series expansion (A5.28) with each term modified by the corresponding Lanczos damping factor (A5.71). Use a value of M such as $M = 10$. A good idea is to write a small computer or calculator program so that you can investigate the dependence on M. For example, I found that a major effect of the damping factors is that a much larger value of M is required to get a fit that looks as good as when damping is omitted.

(*c*) Show that with Lanczos damping the σ_k factors (A5.71) obtained by the procedure given in (*a*) are independent of the function whose Fourier series is being obtained.

[A5.5] {2}   (Dirac delta function, named after P.A.M. Dirac a famous scientist in quantum theory)

(*a*) In the Fourier integral transform of the triangle function in A5.7, consider the limit as the width T of the function tends to zero. Thus, the triangle becomes a very sharp and tall spike having unit area. Use the results of Exercise A5.7.4 to show that the Fourier integral transform $Y(\omega)$ becomes very spread out around $\omega = 0$, and that the subsidiary maxima become negligible.

(*b*) Suppose that in this limit of very small width $y(t)$ represents a sudden impulse, such as a mechanical blow or a line-voltage spike. Describe the frequency response, and give an intuitive explanation of it.

[A5.6] {2}   (FFT in remote-sensing applications)

A satellite takes high-resolution images of Earth's surface. Each image scans $3°$ of latitude and $3°$ of longitude. Assume that the Nyquist criterion in A5.2 limits the spatial detail in a discrete Fourier transform analysis. To simplify the calculations, make them for a satellite in equatorial orbit.

(*a*) Estimate the linear spatial resolution Δ of the satellite images if $N = 1024$ is used. The radius of Earth is $r_E = 6370$ km.

(*b*) If croplands are scanned by this satellite imaging system, show that the minimum area of a given crop that can be resolved is about 10 hectare, assuming a square image with equal resolution in each direction. (A hectare is the area of a square 100 m on a side.)

(*c*) From your knowledge of world agricultural practices, discuss how reliable the results obtained using similar satellite imaging systems for the total cultivated area of different crops would be when data for the Orient are assembled relative to data for the North American Midwest.

The reference list for space vehicles and satellites in L11 indicates several sources on FFT applications to remote sensing.

References on Fourier expansions

Abramowitz, M, and I. A. Stegun, *Handbook of Mathematical Functions*, Dover, New York, 1964.

Bennett, W. R., *Scientific and Engineering Problem Solving with the Computer*, Prentice-Hall, Englewood Cliffs, N.J., 1976.

Brigham, E. O., *The Fast Fourier Transform*, Prentice-Hall, Englewood Cliffs, N.J., 1974.

Brillouin, L., *Science and Information Theory*, Academic Press, New York, 1956.

Churchill, R. V., *Fourier Series and Boundary Value Problems*, McGraw-Hill, New York, 1963.

French, A. P., *Vibrations and Waves*, W. W. Norton, New York, 1971.

Herival, J., *Joseph Fourier; The Man and the Physicist*, Oxford, New York, 1975.

Oppenheim, A. V., and R. W. Schafer, *Digital Signal Processing*, Prentice-Hall, Englewood Cliffs, N.J., 1975.

Otnes, R. K., and L. Enochson, *Applied Time Series Analysis*, Wiley, New York, 1978.

Papoulis, A., *Signal Analysis*, McGraw-Hill, New York, 1977.

Pippard, A. B., *The Physics of Vibration*, Vol.1, Cambridge University Press, 1978.

Pratt, W. K., *Digital Image Processing*, Wiley, New York, 1978.

Wolf, K. B., *Integral Transforms in Science and Engineering*, Plenum Press, New York, 1979.

Wylie, C. R., *Advanced Engineering Mathematics*, McGraw-Hill, New York, 1975.

A6   INTRODUCTION TO DIFFERENTIAL EQUATIONS

In this chapter we will develop practice in setting up and solving differential equations, predominantly as they describe physical systems. However, the applications will be broad; how student scores are related to effort (in A6.2), the kinematics of world-record sprints (in A6.3), the exponential growth of consumption by the United States in the twentieth century (in A6.4), and the description of systems in nature having self-constraints (in A6.5). Numerical methods for first-order differential equations are introduced in A6.6.

A wide selection of problems is offered at the end of the chapter. These extensive problems cover a broad range of applied science; world-record sprints, solar-energy collectors, doubling-times for resource use in the United States, the use of decibel scales in noise measurement and fiber-optics communications, population dynamics, competition between warring armies, and the numerical solution of first-order differential equations.

First-order linear differential equations is the main topic of this chapter. Second-order differential equations are covered in A7, which contains further examples and numerical methods.

From the particular to the general

We develop the methods of solution of differential equations from the particular to the general. That is, a scientifically interesting problem will be cast into differential equation form, and this particular differential equation will be solved as part of deriving the method of solution. Then the more general method of solution for this class of differential equation will be investigated. I hope that this order of presentation will motivate you better than the reverse order, typical (and perhaps desirable) in mathematics texts, but quite untypical in applied science.

A6.1 {1}   Differential equations and physical systems

In this section we study several examples of how differential equations arise in formulating and solving problems in the applied sciences. In pure mathematics the emphasis in studying differential equations is often on the existence and the general nature of solutions. For the worker in applied science it is most important to recognize when a problem can be put into differential equation form, to solve this differential equation, to interpret the quantities that appear in it, and to relate the solutions to data and observations.

Why differential equations?

Differential equations, relating rates of change for infinitesimal changes of independent variables, are very commonly used to describe physical systems when the phenomena to which such equations apply are assumed to vary in a continuous way. Historically, differential and integral calculus have been applied for three centuries to the areas of mechanics and electromagnetism, beginning with the invention by Isaac Newton (1642–1727) and Gottfried Leibniz (1646–1716) of the differential calculus. Nineteenth-century research in these areas focused on electricity and magnetism, and the twentieth century has seen the development of electronics, electromagnetic communications, and microelectronics. Both the principles and the practical devices derived from them require for their understanding and design extensive use of differential equations.

Formally, a *differential equation* is an equation involving one or more derivatives of a function, say $y(x)$, and an independent variable x. To have obtained a solution to the differential equation is to have produced a relation between y and x that is free of derivatives and that gives y explicitly as a function of x.

In general, there will be more than one such relation for a given differential equation. For example, there is an infinite number of straight lines having the same slope (first derivative) but different intercepts. The choice of the appropriate relation will depend strongly on constraints on the solution. (In the example of the straight line with given slope, the constraint is the value of the intercept.) These constraints are called *boundary values* or *initial conditions* and must be specified independently of the differential equation. Indeed, knowledge of these constraints for the particular problem at hand is often a guide when solving the differential equation. This is illustrated frequently in the examples that follow.

▶ *Exercise* A6.1.1   Suppose that $y(x) = c_1 \exp(x) + c_2 \exp(-x)$, where c_1 and c_2 are constants.
(a) What choice of c_1 and c_2 is necessary in order that y remain finite as $x \to \infty$?
(b) Is the condition $y'(0) < 0$ sufficient to make y finite for all positive x? If not, determine a sufficient condition. ☐

Time is most often the independent variable appearing in the examples of differential equations studied in this chapter. This probably arises from the predictive capability obtained by solving a differential equation in the time variable. For example, from current birth and mortality rates of humans we can predict (hopefully incorrectly) when there will be standing room only on Earth. The present disintegration rate of a ^{14}C radionuclide sample can be used to estimate when the plant tissue that assimilated it died, as considered in detail in computing laboratory L8.3.

Notation and classification

The notation for derivatives used in this chapter is as follows: The quantity $D_x!!y$ is the first derivative of y with respect to tariable x, and $D_x y$ is the first derivative of y with respect to variable x, and $D_x^2 y$ is the corresponding second derivative, notations preferable to dy/dx with its suggestion of division, a notation useful only in numerical applications, as in A4.3. The D_x notation is also useful when the powerful concept of a differential operator is introduced in higher mathematics. To develop a technical vocabulary of terms relating to differential equations, we consider their classification, as follows:

There are two types of differential equations; those involving total derivatives (such as D_x) are called *ordinary differential equations*, while those involving partial derivatives (such as $\partial y/\partial x$) are termed *partial differential equations*. We consider only ordinary differential equations in this book. This greatly simplifies both the analytical and numerical work. Furthermore, one of the standard methods of solving partial differential equations is to reduce them to ordinary differential equations.

If the derivative is taken n times, as in D_x^n, then the differential equation is said to be of nth *order*. In mechanics, for example, if p is a momentum component, t is time, and F is the corresponding force component, then Newton's equation, written as $D_t p = F$ is first order in t. The displacement component, say x, and the mass m are related to F by $mD_t^2 x = F$, which is a second-order differential equation in the variable t.

▶ *Exercise* A6.1.2
(*a*) Write Newton's force equation for a single component as a pair of first-order differential equations.
(*b*) Show how any nth order differential equation can be written in terms of n first-order differential equations. (This result is important for numerical methods of solving differential equations. See sections A6.6, A7.5, and Chapter L11.) □

If the derivative appears raised to an algebraic power, $(D_x y)^m$, in a differential equation, then the equation is of mth degree. Generally, $m = 1$, (*linear* differential equations) are preferred, because those of higher degree usually require the simultaneous solution of algebraic and differential equations. In this text we consider mainly ordinary first-, and second-order linear differential equations.

Homogeneous and linear equations

There are two distinct definitions of ''*homogeneous*'' as applied to differential equations. The first definition is that a differential equation is homogeneous if it remains unchanged when both x and y variables are multiplied by the same, nonzero, scale factor. For example, $D_y = x \sin(y/x)$ is homogeneous in this sense.

The other definition of a homogeneous differential equation, usually applied to linear ordinary differential equations, is that an equation is homogeneous if constant multiples of solutions are also solutions. For example, $D_t^2 y = -ky$, the differential equation for simple harmonic motion, is a homogeneous differential equation.

The power of such linear differential equations is that their solutions are additive. We speak of the *linear superposition* of solutions of such equations. Almost all the fundamental equations of physics, chemistry and engineering, such as Maxwell's, Schroedinger's, and Dirac's equations, are homogeneous linear differential equations.

A physical device that often behaves analogously to a linear differential equation is the "black box" used in electronics (rather than in magic shows). Such a device is termed linear if its output is proportional to the input. For example, in signal detection the primary amplifiers are usually designed to be linear, so that the weak and strong signals are increased in the same proportion. A black-box Fourier analyzer, either as hardware in a very-large-scale-integration (VLSI) chip (of the kind discussed in the article by Thompson) or as software in a FFT program (see A5.8 and L9.2), is a linear device. Lasers, on the other hand, operate as highly nonlinear devices.

► *Exercise* A6.1.3

(*a*) Give a scientific explanation of why "black-box" electrical equipment is often colored black.

(*b*) Why do magicians often use devices, draperies, and clothing that are black? □

A6.2 {1}   Separable differential equations

One type of differential equation that occurs frequently in applied science and that can be solved readily is the separable differential equation. For these equations the variables can be separated so that the derivatives can be integrated separately. This is exemplified in the following differential equation based on a decade of my students' experiences in the courses on which this book is based.

Relating student scores and work

The problem is as follows: An instructor awards a score S that increases relative to the student's work, W, such that the increase in score for a given amount of extra work is proportional to the difference between the maximum score S_m and the current score S. Thus, it takes students about 6 times as much work to increase their scores from 90% ($S_m - S = 10\%$) to 95% than to move from 40% ($S_m - S = 60\%$) to 45%. Let us set up the appropriate differential equation and solve it to find a formula that will enable S to be calculated for given W.

Denote a finite increase in work by ΔW, and the corresponding increase in score by ΔS. Now we can write $\Delta S/\Delta W = (S_m - S)/\tau$, where τ is a constant of proportionality. Its choice will become clear as you work a little more into this text. Assuming that the work increment ΔW can become infinitesimal (as justified by experience), we take the limit of the left-hand side of this equation and obtain the differential equation relating score to work

$$D_w S = (S_m - S)/\tau. \tag{A6.1}$$

To solve this differential equation, note that both sides can be divided by $(S_m - S)$, then integrated with respect to dW, to obtain

$$\int dW \, D_w S/(S_m - S) = \int dW/\tau. \tag{A6.2}$$

Both the integrals are indefinite integrals. Under the left-hand integral we can replace $dW \, D_w S$ by dS, so that the integral becomes $\int dS/(S_m - S)$, which is just $-\ln (S_m - S)$, as can be verified by differentiating this result. The right-hand integral is just $W/\tau + constant$. On equating the two sides, then taking exponentials, we get

$$S_m - S(W) = C \exp (- W/\tau), \tag{A6.3}$$

where C is another constant.

Appropriate boundary conditions on the score-work differential equation solution lead to determination of C, as follows. A reasonable initial condition relating the score to the work is that no work produces no score; that is $S(0) = 0$. Inserting this in (A6.3) gives $C = S_m$. On rearranging (A6.3), the final solution to the differential equation relating student score S to work W is obtained as

$$S(W) = S_m[1 - \exp (- W/\tau)]. \tag{A6.4}$$

▶ *Exercise* A6.2.1   Carry out in detail the steps leading from the differential equation (A6.1) to the result (A6.4) for the relation between student work and score. □

The relation between S and W given by equation (A6.4) is shown graphically in Figure A6.2.1. The interpretation of τ is obtained through the following exercise.

▶ *Exercise* A6.2.2
(*a*) Show that if a student does an amount of work $W = \tau$, then 63% of the maximum score is attained, at least according to (A6.4).
(*b*) Assume that W is measured in hours. Make a graph of $S(W)/S_m$ versus W for an acceptable range of W (positive, of course). Use $\tau = 1$ hr. □

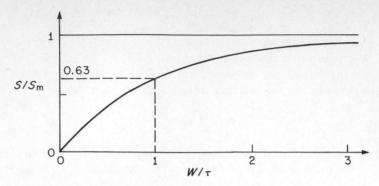

Figure A6.2.1   Student score relative to the maximum score, plotted against work in units of τ, according to (A6.4).

Suppose, for example, that W is measured by the study time in hours, then τ is the number of hours of work required to get the 63% score. For reasons that are not clear from this example, τ is called the "relaxation time". However, this terminology will become clearer in later examples on differential equations involving exponentials.

▶ *Exercise* A6.2.3   (More work on the student scores equation)
(a) From data on the relation between S and W available from your own experience with a particular instructor and course, determine your relaxation time, or work-performance index, τ, for the course. One way of doing this is to take your scores from several assignments and to make estimates of the work you expended on each. The time τ can be estimated for each assignment by solving (A6.4) for τ, given $S(W)$ and a guess of W. The τ estimates from each assignment can then be averaged.
(b) As a more elegant way of estimating your personal value of τ, transform variables in (A6.4) so that W appears linearly in the equation. Use the transformed equation to make a straight-line least-squares fit, as described in A4.7 and in the computer program in L8, in order to extract a best-fit value of τ.
(c) Why is the method of determining τ suggested in (b) better than that in (a)?
(d) Under what conditions would you be justified in using different weights for different assignments in (b)? ☐

Generalizing separable equations

From the previous example we can generalize the method of solution of first-order separable differential equations. Given

$$D_x y = f(x)g(y),   \text{(A6.5)}$$

in which the right side is indicated as being separated into the product of a function of x with a function of y, the solution of the differential equation reduces to the problem of performing the two separate integrals in

$$\int dy/g(y) = \int dx\, f(x) + constant. \tag{A6.6}$$

The constant is determined by the initial or boundary conditions. In the above score/work example we have $x = W$, $y = S$, $f(x) = 1/\tau$ and $g(y) = S_m - S$.

A6.3 {2}   First-order linear equations: World-record sprints

In this section we set up a realistic problem from athletics in which a first-order linear differential equation arises. Our analysis is phenomenological in that it is concerned only with the kinematics involved in competitive sprinting and does not investigate the physiological and psychological factors that govern such athletic contests. Therefore, in the language of circuit theory and modeling, it is a *lumped-parameter* analysis.

Kinematics of world-record sprints

According to J. B. Keller's theory of competitive running, published in 1973, in track sprints the speed v of a world-class runner in races of up to 300 meter distance can be described by the following formula for the acceleration

$$D_t v = A - v(t)/\tau. \tag{A6.7}$$

As of 1973, the appropriate constants were the acceleration parameter $A = 12.2$ m/sec^2 and the "relaxation time" $\tau = 0.892$ sec. (In Keller's paper F rather than A is used, but F is easily confused with force.) In this differential equation the net acceleration is determined by the driving acceleration A minus a resistance proportional to the instantaneous speed $v(t)$ at time t. The physiological justification of this equation, and its restrictions, are discussed by Keller, by Ryder, and by Alexandrov and Lucht.

   To solve the differential equation for world-record sprints with the appropriate initial conditions, we use our familiarity with athletics and a little mathematics to build up the appropriate solution from limiting cases.

   In track sprints the runners start from rest, so that their speed $v = 0$ at $t = 0$. Thus, near the start of the race a good approximation to the differential equation will be $D_t<u><u>v \approx A$, which has solution $v(t) = At$. On the other hand, suppose (slightly unrealistically) that the A term on the right-hand side of the differential equation were negligible, then $D_t<u><u>v \approx -v(t)/\tau$. This has the solution $v(t) = C \exp(-t/\tau)$, where C is a constant, as can be verified by differentiation. However, it can not be exactly right because it predicts that v is always nonzero, but we are sure that $v(0) = 0$. (Don't jump the starting gun.)

Now that we have two limiting (but slightly incorrect) solutions, is there a solution consistent with these? If we extend the second solution to the limit $t/\tau \ll 1$, then it predicts $v(t) \approx C(1 - t/\tau)$, in which we have used the first two terms in the Maclaurin expansion of the exponential. For this to give the first approximate solution for small t, we would need to add a constant $-C$ to it and to have $-C/\tau = A$, that is, $C = -A\tau$. This fix produces a patched-up solution to the sprinter problem

$$v(t) = -A\tau \exp(-t/\tau) - (-A\tau). \tag{A6.8}$$

▶ *Exercise* A6.3.1   As a check on this method of deriving the solution, verify that it satisfies the original differential equation (A6.7) by performing the indicated differentiation. □

Thus, writing out the solution to the differential equation for world-record sprinters neatly, we have their speed as a function of time

$$v(t) = A\tau[1 - \exp(-t/\tau)]. \tag{A6.9}$$

Now that we have derived the relation (A6.9) between speed v and time t in world-record sprints, it is interesting to investigate some of its properties.

Limbering up

You may have recognized that our sprinter differential equation is also a separable equation, formally (at least) similar to the student score/work problem in A6.2. If so, get some more exercise with the following.

▶ *Exercise* A6.3.2   Use the method of separating the differential equation, as in A6.2, to solve the speed-time relation for world-record sprints, starting with (A6.7). □

To gain additional insight into the kinematics of sprinting, and incidental insight into solving differential equations, consider the following exercise.

▶ *Exercise* A6.3.3   (Distance against time in the Keller model)
(*a*) Set up and solve the differential equation that gives the distance $D(T)$ run in a total time T. Assume that the runner starts from rest. Show that

$$D(T) = A\tau\{T - \tau[1 - \exp(-T/\tau)]\}. \tag{A6.10}$$

(*b*) Write and run a program for pocket calculator or computer to evaluate $D(T)$ for

the current world-record times for 50, 100, and 200 meters. Calculate the average percentage discrepancy between Keller's model and the current data.
(*c*) Compare the initial acceleration of world-class sprinters with that of fast sports cars and of other large mammals.
(*d*) Predict the time to run 40 km (approximately a marathon) according to the $D(T)$ equation. Why is this prediction unrealistic? □

Figure A6.3.1 shows the fit of the distance-time relation (A6.10) to the 1973 world records. Problem [A6.1] will further your understanding of the parameters A and τ, while Problem [A6.2] suggests that you compare the coefficients in Keller's formula between world-class male and female sprinters.

Physics and physical activity

Since Keller's study of the kinematics of sprints there have been many discussions of the kinematics of this and other athletic events. McMahon and Greene studied the effect of track surface resiliency on running style and analyzed this with computer programs. Alexandrov and Lucht showed that the greater centripetal acceleration required by the inside runner compared with the outside runner on curved tracks gives the inside runner a nonnegligible mechanical disadvantage. However, the psychological stimulus provided to the inside runner by starting from behind may provide the necessary impetus. A similar analysis is presented in the article by Neie. Athletic records and their relation to human endurance are the subject of an article by Riegel in *American Scientist*.

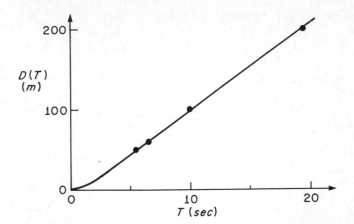

Figure A6.3.1   Distance-time relations for world-record sprints, according to (A6.10), with data for 1973 records. Note that the curve is non-linear only in the first few seconds.

In other sports, the forces involved in the marital arts (such as karate) are described in an article by Blum. The very large impact forces in karate are the subject of a *Scientific American* article by Feld, McNair, and Wilk.

General first-order differential equation

We now tackle (in the nonathletic sense) the more general differential equation

$$D_x y = Q(x) - P(x)y, \tag{A6.11}$$

where $Q(x)$ and $P(x)$ are arbitrary, but fairly well-behaved, functions of x. For example, in the sprinting problem we have $P(x) = 1/\tau$ and $Q(x) = A$, both functions independent of x, which is why the solution was straightforward.

Many scientifically interesting problems can be cast into this differential equation form. The function $Q(x)$ is a "driving" or "source" term that determines initially the rate at which y is changed by x. The function $P(x)$ which multiplies y is typically a feedback term, since it produces a rate of change in y proportional to the current value of y and scaled by the value $P(x)$.

To solve the general equation, we try to arrange that all the y dependence is isolated into a single term and that this term is a straightforward derivative. (Clearly you have to have seen the answer somewhere else!) Write the differential equation as

$$\varrho D_x y + \varrho P y = \varrho Q, \tag{A6.12}$$

where ϱ is to be chosen to achieve our goals. Note that each of ϱ, y, P, and Q depend implicitly on x, but to help the typesetter (WJT) we omit this dependence. Can the left side be made a perfect derivative of x? This requires

$$\varrho D_x y + \varrho P y = D_x(\varrho y). \tag{A6.13}$$

On expanding the right side, we find that this further requires

$$D_x \varrho = \varrho P. \tag{A6.14}$$

▶ *Exercise* A6.3.4   Integrate this differential equation by separation of variables to show that

$$\varrho(x) = \exp\left[\int dx'\, P(x')\right]. \tag{A6.15}$$

☐

This is a general prescription for $\varrho(x)$, which is called the *integrating factor*. The original differential equation is now transformed into

$$D_x(\varrho y) = \varrho Q. \tag{A6.16}$$

If we now integrate both sides with respect to x, we find the final solution to the general first-order linear differential equation

$$y(x) = [\int dx'' \, \varrho(x'')Q(x'') + c]/\varrho(x), \tag{A6.17}$$

where c is a constant of integration.

▶ *Exercise* A6.3.5   Show the preceding steps in solving the general first-order equation in detail. □

Here is another check on the successful solution of this problem.

▶ *Exercise* A6.3.6   Since one should verify a solution by substitution into the original differential equation, do this for solution (A6.17) inserted in (A6.11). □

Note that although a general formula for solving the differential equation (A6.11) has been obtained, successful use of (A6.17) still requires that the indicated indefinite integrals can be performed. However, as you will know if you have covered the material on numerical integration (A4.4 and L5), the solution of (A6.17) in numerical (rather than analytical) form can usually be obtained quite directly.

▶ *Exercise* A6.3.7   (World-record sprints from the general solution)
(*a*) Apply formula (A6.11) to the world-record sprints differential equation (A6.7) to show that the integrating factor therein is $\varrho(t') = \exp(t'/\tau)$, and that the solution (A6.17) translates into

$$v(t) = A\tau[1 - (c/A\tau) \exp(-t/\tau)] \tag{A6.18}$$

(*b*) Show that $c/A\tau = 1$ if $v = 0$ at $t = 0$, that is, the runner does not jump the gun. □

After all this athleticism, you would probably like something more relaxing as a diversion.

A6.4 {2}   Diversion: The logarithmic century

In his book, *The Logarithmic Century*, Ralph E. Lapp surveys production and consumption in the United States from the Industrial Revolution through the early 1970s. In almost every graph in this book a semilogarithmic scale for consumption versus time is needed to fit the book page. That is, consumption of many products in the United States was roughly exponential with time during that era. For example, horse power declined while horsepower increased at a rate indicated in Figure A6.4.1.

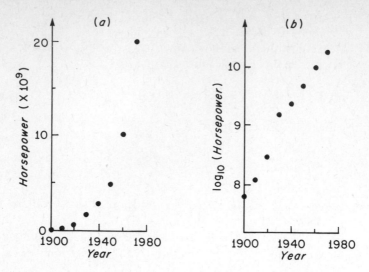

Figure A6.4.1   Horsepower growth in the United States since 1900. Curve (a) is a linear display, and (b) is a logarithmic display. Note the effects of automobile mass production in the 1920s and of the Great Depression in the 1930s.

To relate exponential growth to differential equations, note that if the amount present is $N(t)$ at time t, and if this increases exponentially with time, then

$$N(t) = N(0) \exp (\lambda t), \qquad (A6.19)$$

where $N(0)$ is the amount initially present and λ is a constant. On differentiating both sides with respect to t we find the differential equation

$$D_t N = \lambda N(0) \exp (\lambda t) = \lambda N(t). \qquad (A6.20)$$

This last equality can be transformed to give

$$D_t N(t)/N(t) = \lambda. \qquad A6.21)$$

Thus, exponential change implies instantaneous *fractional* rates of change that are constant with time. Note that the exponent constant λ can be positive (for growth) or negative (for decay).

▶ *Exercise* A6.4.1   Verify all the intermediate steps between equations (A6.19) and (A6.21) for exponential behavior. □

▶ *Exercise* A6.4.2   As (A6.19) is written, $N(t)$ seems to depend on the choice of time origin. Show, however, that a more general result is

$$N(t) = N(t_1) \exp [\lambda(t - t_1)], \tag{A6.22}$$

in which the time t_1 is arbitrary. Do this in two ways; first by verifying that this result satisfies the differential equation (A6.21), and second by deriving it from (A6.19). □

This exercise shows that the exponential model is memoryless in the sense that ratios of N values depend only on the elapsed time between when they are measured. If such a model were applied to predict computer reliability, a computer system that has not failed in the past 50 hours is equally likely not to fail in the next 50 hours. Does this agree with your experience?

Interpreting exponential behavior

The interpretation of exponential change with time is governed mainly by the parameter λ. If λ is real and *negative*, then N decreases with time. This is characteristic of the radioactive decay of atomic nuclei, for which $-\lambda$ is the fractional rate of decay per unit time. (The study of radioactivity by other means, namely Monte-Carlo simulation, is made in computing laboratory L6.) The depletion of natural resources that are not readily replaced, such as petroleum and tigers, may often be similarly modeled, except perhaps when N becomes relatively small. In studying the relaxation of an initially excited physical system, such as the current in a simple electrical circuit or damped mechanical vibrations, the decay is often exponential, as discussed in A7.4. The decrease of intensity of radiation as it penetrates an absorbing medium is often approximated by exponential decay with depth of penetration. This example is applied to the design of solar energy collectors in Problem [A6.3].

If λ is real and *positive*, then N is predicted to increase indefinitely with time. The value of N is then appropriately displayed on semilogarithmic scale, since from (A6.19)

$$\ln [N(t)] = \ln [N(0)] + \lambda t. \tag{A6.23}$$

This equation is linear in the time t, with slope the growth constant λ and intercept the natural logarithm of the number initially present, $N(0)$. On a base-10 logarithmic scale (A6.23) is changed to

$$\lg [N(t)] = \lg [N(0)] + (\lambda/2.30258)t, \tag{A6.24}$$

in which the divisor is $\ln (10)$. Examples of this exponential growth are given in Lapp's book and are the subject of Problem [A6.4].

The transformation from the exponential form to the logarithmic form enables the use of straight-line least-squares fitting, as discussed in A4.7 and in computing laboratory L8. Such a logarithmic transformation is also made in L10.3 as part of the Fourier-transform analysis of resonance line widths.

Bell's decibels

The attenuation of signals is often expressed in terms of decibels (dB). The unit of 1 bel = 10 decibels is named after Alexander Bell (1847–1922), an American audiologist and the inventor of the telephone. Two signals of intensities I_1 and I_2 have a decibel difference of

$$D = 10 \lg_{10} (I_1/I_2). \tag{A6.25}$$

The abbreviation for decibel is dB.

Decibel scales are frequently used in physiology because the ear and the eye are approximately logarithmic detectors over much of their range of sensitivity. That is, the observed difference of loudness or brightness of two sources depends mainly on the decibel difference between them, rather than on the absolute difference, or the ratio, of their intensities.

▶ *Exercise* A6.4.3   What are the survival advantages of sense organs responding logarithmically? What are the disadvantages? □

The use of decibel scales in medicine and physiology is discussed in Chapters 13 and 15 and in Appendix A of Cameron and Skofronick.

The following examples illustrate use of logarithmic scales and of decibels. If a signal intensity attenuates exponentially with respect to variable x (where x is typically distance, time, or frequency), then the decibel level D satisfies

$$D_x D = constant, \tag{A6.26}$$

that is, $D \propto x$.

▶ *Exercise* A6.4.4   The human ear has a threshold of hearing of about 10^{-12} W/m² sound intensity. Speech is typically at 60 dB above threshold, and aural pain usually begins at about 120 dB above threshold (jet engine noise close up).
(*a*) Express these dB values as intensities.
(*b*) I have noticed that airport ground crews use ear mufflers about 2 cm thick. Assuming that this reduces sound intensity from painful to speech level, calculate how much thicker the muffler should be to reduce the sound level to a whisper (10^{-10} W/m²). □

Further examples on the use of decibel scales are suggested in Problem [A6.5].

A6.5 {2}   Nonlinear differential equations

If the dependent variable appears in a differential equation to a power other than linear, then we have a nonlinear differential equation. So far in this chapter the examples have all been linear equations. Non-linear equations are generally difficult to handle because one must handle nonlinear algebraic equations at the same time as puzzling over the differential calculus aspects.

In this section we consider a relatively straightforward nonlinear differential equation, that for the logistic growth curve. Several examples of nonlinear equations are given in the book by Dym and Ivey on mathematical modeling.

The logistic growth curve

In A6.4 the exponential decay or growth of a system was discussed in terms of its differential equation. To be realistic, exponential growth can not continue indefinitely. In most systems that begin with exponentially increasing behavior, inhibitory effects that increase as $N(t)$ increases are expected. For example, in an electronic circuit this could be obtained by having negative feedback proportional to the output. In the population dynamics of natural species there are often self- or external constraints that limit the rate of growth of the species, as discussed in Problem [A6.6].

An interesting differential equation that models such inhibitory effects is the nonlinear differential equation

$$D_t N(t) = \lambda N(t) - \lambda_1 N^2(t), \tag{A6.27}$$

with both λ and λ_1 positive. In equation (A6.27) the inhibitory effect on the number present at time t, $N(t)$, is proportional to the square of the current value $N(t)$.

An equilibrium value of $N(t)$ is attained if there is a solution of this equation such that $D_t N(t) = 0$. Setting the right-hand side of (A6.27) to zero gives $N_m = \lambda/\lambda_1$ for the equilibrium value N_m to which $N(t)$ tends after a long time. Thus, we can rewrite equation (A6.27) as

$$D_t N(t) = \lambda N(t) [1 - N(t)/N_m]. \tag{A6.28}$$

To solve this differential equation, note that for times small enough that $N(t)/N_m \ll 1$ we have the original exponential-growth equation. After these early times the slope $D_t N$ decreases to zero as $N(t)$ approaches N_m.

▶ *Exercise* A6.5.1 (Solution of the logistic growth equation) Assume a solution of the logistic growth differential equation (A6.27) of the form

$$N(t) = A/[1 + B \exp(-\lambda t)], \tag{A6.29}$$

which is consistent with the observations made about (A6.28).
(a) By using the immediately preceding arguments, show that $A = N_m$.
(b) Show that the assumed form of the solution satisfies the original differential equation for any value of B.

(c) If the number initially present is $N(0)$ show by substitution into (A6.29) that $B = N_m/N(0) - 1$. □

On assembling all the results from this exercise, we find that the solution of the differential equation for a quadratically self-inhibiting system has the number present at time t, $N(t)$, given by

$$N(t) = N_m/[1 + (N_m/N(0) - 1) \exp(-\lambda t)], \tag{A6.30}$$

in which N_m is the equilibrium value of the population N, and $N(0)$ is the initial population. The curve of $N(t)$ versus t is called the *logistic growth curve*. It is shown in conveniently scaled form in Figure A6.5.1.

▶ *Exercise* A6.5.2   As an alternative (and very direct) method of solving the logistic growth equation (A6.27) cast it into the separable equation mold as in A6.2. Then set $n(t) = N(t)/N_m$, and show that $D_t n/[n(1 - n)] = \lambda$. Use the identity $1/[n(1 - n)] = 1/n + 1/(1 - n)$ to integrate this differential equation with respect to n in order to obtain

$$\ln(n) - \ln(1 - n) = \lambda t + c, \tag{A6.31}$$

where c is a constant of integration. Thence, by taking exponentials, derive the logistic growth curve (A6.30). This method of solution is also discussed in section 39 of Haberman's text on mathematical modeling. □

Exploring logistic growth

The curves of logistic growth have several interesting properties, which are not self-evident, mainly because of the nonlinear nature of the differential equation (A6.27). The following exercises give you opportunities to explore the properties of logistic-growth curves. These properties are especially important in modeling population dynamics in ecology, harvesting of biological natural resources, and in the control of epidemics.

▶ *Exercise* A6.5.3   (Properties of the logistic growth curve.) It is convenient to produce changes of scale in both N and t so that a single parameter, the ratio of initial to equilibrium populations, $N(0)/N_m$, determines the curve.
(a) Verify that if the logistic growth curve (A6.30) is scaled so that $n(t') = N(t)/N_m$, with $t' = \lambda t$, then it can be written

$$n(t') = 1/[1 + (1/n(0) - 1) \exp(-t')]. \tag{A6.32}$$

Note that $n(0) = N(0)/N_m$ is just the ratio of initial to equilibrium populations.
(b) Show that $n(t')$ satisfies the differential equation

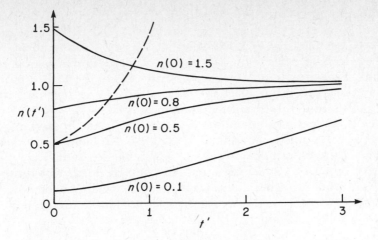

Figure A6.5.1   Logistic growth curves, scaled in both time and population according to (A6.32). The asymptotic values are unity. The dashed curve shows unrestrained exponential growth.

$$D_t n(t') = n(t')[1 - n(t')].$$ (A6.33)

(c) Calculate and draw curves of $n(t')$ versus t' for each of several values of $n(0)$, and verify that they are similar to Figure A6.5.1. You may find the plotting procedures from computing lab L4 useful for this.

(d) Show that if the logistic growth curve has a point of inflection, then it occurs at $t' = \ln [1/n(0) - 1]$.

(e) Consider the case $n(0) > 1$; that is, the initial population is greater than the equilibrium population. Sketch the resulting decrease of the population with time, and show that there is no point of inflexion. ☐

Even though growth is limited for the logistic growth curve, as formula (A6.32), Figure A6.5.1 and your own graphs from Exercise A6.5.3 clearly show, the following exercise may convince you that prediction of the final population N_m is not straightforward.

► *Exercise* A6.5.4   (Predicting equilibrium population)
(a) Use the differential equation (A6.27) to show that N_m can be predicted from the populations N_1, N_2 at times t_1, t_2, and the growth rates $N'_1 = D_t N$ at $t = t_1$ and $N'_2 = D_t N$ at $t = t_2$, according to

$$N_m = (N_1 N'_2/N_2 - N_2 N'_1/N_1)/(N'_2/N_2 - N'_1/N_1).$$ (A6.34)

(b) Why is an accurate value of the equilibrium population N_m difficult to obtain using this equation applied to empirical data with their associated errors? ☐

Applications of the logistic growth differential equation to bioeconomics and mathematical modeling of biological systems are given extensive treatments in the texts by Clark and by Gold. Problem [A6.6] considers modification of this equation to include effects of harvesting or predation. Several practical applications to the growth of biological systems and to the spread of technological innovations are given in chapters 5 and 6 of the text on differential equation models edited by Braun, Coleman, and Drew.

Interspecies competition among humans, namely that in warfare, leads to consideration of coupled differential equations. Problem [A6.7] considers Lanchester's Law of military survival, which models attrition in warfare between two opposing armies as a pair of coupled linear differential equations. The quantitative aspects of a wide range of social phenomena, some of which are treated by the methods of statistical mechanics, are presented in the book by Montroll and Badger.

A6.6 {2}　Numerical methods for first-order equations

As we discovered in the preceding sections, the analytical solution of a differential equation, if possible, requires techniques specific to each differential equation. On the other hand, numerical methods for differential equations can often be applied to a wide variety of problems. But they are not foolproof. Therefore, experience in identifying, programming, and checking appropriate numerical methods for the solution of differential equations are all important.

In this section we emphasize the basic principles and the formal developments, while several of the exercises suggest numerical trials. The development is continued in A7.6 and A7.7 for second-order equations, whereas the integration of the equations of motion for space-vehicle motions by the numerical techniques derived in this section and in A7 is part of computing lab L11.

A differential equation relates the derivatives of a dependent variable $y(x)$ to a function $f(x,y)$. The process of solving the differential equation then consists of extrapolating the values of y using information about the derivatives and about f. Alternatively, since $y(x) = \int dx\, y'(x)$, we may consider the solution of a differential equation as an integration process. Thus, one often speaks of "integrating the differential equation."

In this and succeeding sections we will abbreviate derivatives as $y'(x) = D_x y(x)$, $y(x) = D_x^2 y(x)$, etc. for convenience or for clarity.

Predictor formulas

We consider the numerical solution methods in which previous and current values are used to predict y values at nearby x values. The general first-order ordinary differential equation can be written

$$y'(x) = f(x,y).$$ (A6.35)

Suppose that y is known for a succession of x values, equally spaced by h, up to x_k. There are two basic ways of numerically advancing the solution from x_k to $x_{k+1} = x_k + h$. The first way is to approximate the derivative on the left-hand side as in A4.3, using (for example) the forward difference formula

$$y'_k \approx D_+(h) = (y_{k+1} - y_k)/h.$$ (A6.36)

Here we have made the abbreviation $y_k = y(x_k)$, $y'_k = y'(x_k)$, etc. With h suitably small we may get a good approximation to y at the next x value by inserting this in (A6.35), to derive

$$y_{k+1} \approx y_k + hf(x_k,y_k).$$ (A6.37)

This extrapolation, or *predictor*, formula is called *Euler's formula*. It is illustrated in Figure A6.6.1.

▶ *Exercise* A6.6.1   In A4.3 we argued that the central-difference derivative is much more accurate than the forward-difference derivative. Show that the central-difference predictor formula yields

$$y_{k+1} \approx y_{k-1} + 2hf(x_k,y_k).$$ (A6.38)

☐

The second basic method of advancing the solution of (A6.35) is to integrate both sides of it with respect to x

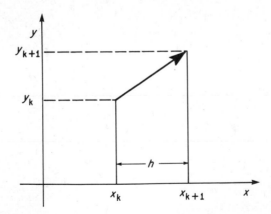

Figure A6.6.1   Euler's predictor formula (A6.37). Note the resemblance to the trapezoid formula in Figure A4.4.1.

$$y_{k+1} - y_k = \int dx \, y'(x) = \int dx \, f(x,y), \tag{A6.39}$$

in which the range of integration is from x_k to x_{k+1}. Although this result is exact, it is not immediately practicable because the unknown y values in the range x_k to x_{k+1} appear under the integral. However, the integral may be approximated by using results from A4.4 on numerical integration.

As an example of integrating the differential equation according to (A6.39), consider approximating the integral from x_k to x_{k+1} by the trapezoid rule (A4.26):

$$\int dx \, f(x,y) \approx [f(x_k,y_k) + f(x_{k+1},y_{k+1})]h/2, \tag{A6.40}$$

which has an error proportional to h^3. If this approximation to the integral in (A6.39) is used, we have an implicit equation for y_{k+1}, since it appears both on the left and right sides of (A6.39). An iterative solution of this equation may be attempted. That is, a trial value of y_{k+1} is used in f to predict y_{k+1} on the left-hand side. We express this as

$$y_{k+1}^{(n+1)} \approx y_k + [f(x_k,y_k) + f(x_{k+1}, y_{k+1}^{(n)})]h/2, \tag{A6.41}$$

for $n \geq 0$. The superscript notation denotes the order of the iteration, and not a derivative (as it does in the Taylor series in A3.3). Note that iterative solution of this equation for y_{k+1} does not guarantee a more correct solution of the original differential equation (A6.35) than that provided by the trapezoidal (or other) integration rule. Formulas in which these integration-iteration methods are used are called *Adams closed formulas*, or *Adams-Moulton formulas*.

▶ *Exercise* A6.6.2   To illustrate use of an improved integration formula in (A6.39), use Simpson's rule (A4.29) applied over the interval x_{k-1} to x_{k+1} with a step size of $2h$ to derive

$$y_{k+1} \approx y_{k-1} + [f(x_{k-1},y_{k-1}) + 4f(x_k,y_k) + f(x_{k+1},y_{k+1})]h/3, \tag{A6.42}$$

which has an error proportional to h^5 only. Note that this is still an implicit equation for y_{k+1} and usually must be solved iteratively. ☐

Both (A6.38) and (A6.42) predict by extrapolation the value of y_{k+1}. It would be desirable to follow the predictor step by a consistent corrector step in which the original differential equation was used to improve the solution accuracy. Such *predictor-corrector* methods are described in, for example, Chapter 8 of Vandergraft and in Hamming's book.

Initial values for solutions

How is the process of solving the differential equation to be started, and how is it to be finished? If we are concerned only with the first part of the question, we have an *initial-value problem*, commonly associated with x as a variable denoting time. If both questions are of concern, then we have a *boundary-value problem*, often associated with x as a spatial variable. (For a physicist in relativity theory such a distinction between time and space is improper, therefore the solutions of relativistic differential equations are rendered more difficult than those that scientists in the slow-moving world use.)

The methods discussed in this section and in A7.6 are most suitable for initial-value problems, to which we restrict our attention. By a suitable choice of x origin, we can assume that the initial values are to be determined at $x = 0$. Thus, we need to specify either y_0 or y'_0, since, according to (A6.35), each determines the other. In Euler's formula, (A6.37), if y_0 is specified, then one can march ahead with y_1, etc; we say that such an algorithm is *self-starting*. For numerically reliable results one should usually start with a step-size h that is no more than half that used for later values of x.

The numerical integration for space-vehicle trajectories in computing laboratory section L11.2 provides an example and exercises in self-starting methods of solving differential equations.

Stability of numerical methods

The stability of numerical methods for solving differential equations is very important, especially if there is little human intervention during evolution of the solution. When computers or calculators are used to do the arithmetic, there should be consistency checking and checkpoint output as the solution is marched forward.

As discussed in A4.2, there are *unstable problems*, and there are *unstable numerical methods* for solving problems that may or may not be unstable. For example, consider the first-order differential equation

$$y'(x) = f(x,y) = ay(x), \tag{A6.43}$$

where the quantity a is constant. This has the analytic solution, examined extensively in A6.4,

$$y(x) = y(0) \exp(ax). \tag{A6.44}$$

This differential equation is fairly unstable if a is small, since for $a = 0$, $y(x)/y(0) = 1$ for all x, but for $a = 0.1$, $y(10)/y(0) = e = 2.7$, and for $a = -0.1$, $y(10)/y(0) = 1/e = 0.37$. Thus a change of only ± 0.1 in the value of a produces the factor 2.7 change in the solution for a not unreasonable value of $x = 10$.

Because of round-off error, at least, an unstable problem will necessarily have unstable numerical solutions. A numerical method for solving a differential equation may be unstable either because the step size h is too large, or because predictor-corrector cycles are imprecise, or for both reasons. For example, consider the preceding differential equation (A6.43) solved by the

Simpson's rule approximation (A6.42), so that, for $f(x,y) = ay$, we have

$$y_{k+1} \approx y_{k-1} + [ay_{k-1} + 4ay_k + ay_{k+1}]h/3, \qquad (A6.45)$$

which can be solved for y_{k+1} by using

$$(1 - b)y_{k+1} - 4by_k - (1 + b)y_{k-1} = 0, \qquad (A6.46)$$

in which $b = ah/3$.

▶ *Exercise* A6.6.3   Assume that this difference equation has a solution of the form $y_k = p^k$, where p is to be determined.
(a) Show that p is determined from a quadratic equation whose two roots are $p_+ = 1 + 3b$ and $p_- = -1 + b$ if $b \ll 1$, that is, if $h \ll 1/a$, as would be natural to require.
(b) Using the results from (a), show that a numerical solution of the differential equation (A6.43) is approximated by the linear superposition

$$y_k \approx A_+(1 + ah)^k + A_-(-1 + ah/3)^k, \qquad (A6.47)$$

where A_+ and A_- are constants. □

The solution (A6.47) to the difference equation can be readily investigated in the limit of small h, as follows. Since $kh = x$, we have

$$y_k(x) \approx A_+(1 + ax/k)^k + A_-(-1)^k(1 - ax/3k)^k. \qquad (A6.48)$$

For $k \gg ax$ the first term approaches $\exp(ax)$, the true and unique solution to $y' = ay$. But there is a second spurious term in the numerical solution, and this term oscillates with k. Its magnitude is of order $\exp(-ax/3)$. If $a > 0$ this is no problem, because the desirable $\exp(ax)$ is dominant and the oscillatory term fades quietly away, like a boojum[1]. However, if $a < 0$ the correct exponential term is shrinking, while the spurious term increases in magnitude and oscillates in sign. Figure A6.6.2 illustrates the stability of the exponential differential equation (A6.43) and its solution (A6.44).

▶ *Exercise* A6.6.4   Show that if $a < 0$ the iteration is stable if performed in the direction of *decreasing* x, that is, in the direction in which y is increasing. How would you establish an initial value in this numerical solution? □

These developments will have convinced you of the importance of checking the stability of numerical methods for solving differential equations. How about trying a numerical example, as suggested in Problem [A6.8]? Section

---

[1]The boojum and its history are discussed by N. D. Mermin, *Physics Today*, April 1981, p.46.

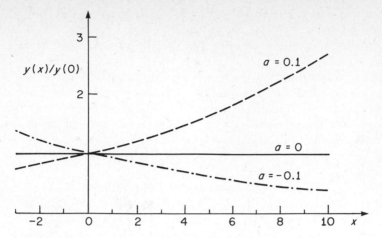

Figure A6.6.2   Stability of the solution of the differential equation (A6.43) for three different values of its parameter.

A7.7 continues this discussion of unstable solutions, applying it to second-order differential equations.

The choice of step size, h, in the differential equation numerical solution formulas was traditionally made by comparing the solutions obtained with step sizes of h and with $h/2$. If the y values obtained did not differ by more than the desired precision, then the halving of h was stopped. Otherwise, the halving process was continued until convergence (not necessarily to the correct results) was reached, until programming bugs were found, until round-off errors became significant, until a deadline was reached, until computing funds were exhausted, or some combination of these truncated the process.

A more efficient scheme is to use information from successive halvings of h to predict values for the limit $h \to 0$. This scheme, called *Romberg integration*, is described in the texts by Hornbeck, Vandergraft, and Acton, and in Chapter 7 of Pizer and Wallace's text.

Problems on first-order differential equations

[A6.1] {1}   (Acceleration and speed in sprints) In investigating the differential equation for world-record sprints we derived in A6.3 that the speed v after time t is modeled by

$$v(t) = A\tau[1 - \exp(-t/\tau)]. \tag{A6.49}$$

Interpretation of this result is aided by deriving the following:
(a) The acceleration of the runner at time t after the start of the race is $a(t) = A \exp(-t/\tau)$.

(*b*) Thus, the maximum acceleration is A, just as the sprinter leaves the blocks. (Since $A = 12.2$ m/sec² $> g = 9.8$ m/sec², the runner has an initial horizontal acceleration significantly greater than the vertical acceleration due to gravity.)

(*c*) A sprinter's acceleration decreases rapidly, and in a race over any of the distances for which Keller's model is appropriate, the acceleration has dropped to $< 10\%$ of its maximum value after less than 3 sec.

(*d*) The runner's final speed during sprints over any of these distances is nearly independent of the total distance run.

[A6.2] {2} (World-class sprinters) In September 1983 the world-record times for sprints were, in pairs (D, T), for men $(100, 9.93)$, $(200, 19.72)$ and for women $(100, 10.79)$, $(200, 21.71)$.

(*a*) In order to see whether athletic performance has improved from $A = 12.2$ m/sec.², $\tau = 0.892$ sec, as obtained by Keller in 1973, solve for A and τ from these two sets of values (segregating the men from the women). One way of doing this is to first eliminate A by taking the ratios of D values, then solving for τ in the resulting equation.

(*b*) Compare world-class men against world-class women in regard to initial impetus and with respect to relaxation of this impetus during the course of a world-class sprint. To do this, compare the A and τ values of male sprinters against those of female sprinters.

[A6.3] {2} (Attenuation in solar collectors) Solar energy collector design requires very efficient transmission of sunlight through the outer wall of the collector. The fraction of the incident intensity I_0 that penetrates a distance x into the collector wall is called the *transmissivity*, $T(x) = I(x)/I_0$. Data for various commercial plastics are as follows:

Material	Thickness (inches) x	Transmissivity $I(x)/I_0$
Teflon	0.001	0.960
Tedlar	0.003	0.910
Lucite	0.080	0.920

(*a*) Assume that the transmission of sunlight through these collector surfaces satisfies the differential equation for the intensity at depth x, $D_x I = -I(x)/L$, where L is a characteristic of each material, called the "extinction length." Show that $T(x) = \exp(-x/L)$.

(*b*) For the thicknesses given in the table, assume that $x/L \ll 1$ and use a Maclaurin expansion of the exponential (as in A4.3) to estimate L for each material.

(c) Determine L for each material by using the exponential form for $T(x)$ and the data in the table. Compare this value with the estimates from (b).

(d) Alternatively, consider the differential equation in (a) as a finite-difference equation $\Delta I/I \approx -\Delta x/L$. Derive L thereby. Why do you get essentially the same values as in the Maclaurin expansion?

(e) What thickness of each material would result in a 15% loss of sunlight?

(f) Most solar energy collectors are made of two or more translucent layers separated by air or partial vacuum. According to the formula for $T(x)$ in (a), for a given total thickness x one gets the same transmission. What other considerations suggest that a multilayer system captures more energy than a single layer of the same total transmitter thickness?

For more on the design of solar energy devices see Meinel and Meinel.

[A6.4]   {1}   (Predictions from exponential growth) The term "doubling time", T_2, is often used to characterize exponential growth. It is the time required for an amount to double from its present amount.

(a) Show that $T_2 = \ln(2)/\lambda = 0.693/\lambda$, independent of the choice of the present time.

(b) The following examples are from Lapp's book, *The Logarithmic Century*.

Growth item	Year	Amount
i. State highway expenditures	1920	$\$4 \times 10^8$
	1970	$\$190 \times 10^8$
ii. Horsepower	1900	20×10^6
	1970	$20{,}000 \times 10^6$
iii. Scientific journals	1700	4
	1950	80,000

Find the exponential-growth constant λ and the doubling time T_2 for each of these examples, assuming exponential growth.

(c) It is often predicted that a major source of energy for the United States, petroleum, will be essentially depleted by the year 2000. Predict for the year 2000 how many horses there would have to be in the United States if all of the horsepower were generated by horses.

(d) Estimate for the year 2000 how many more scientific journals are predicted than for the midcentury, 1950.

(e) Why are the predictions in (c) and (d) probably overpredictions? Suggest alternative energy strategies to reduce the horsepower needed and alternative information technologies to reduce the proliferation of scientific journals.

(f) Computer speeds are doubling about every 4 years. Estimate how much faster computers will be in 2002 than in 1982. Suppose that the fastest computer operations are currently done in 10^{-8} sec. By 2002 what will be the maximum separation of circuit components on which these operations are made, if the operations are not to be significantly hindered by the maximum signal speed, that of light, Assume that a significant hindrance is 10% of the total time.

[A6.5] {2}   (Belles, bells, and decibels)
(*a*) The noise ordinance in Chapel Hill, North Carolina, allows a sound level from fraternity (or sorority) parties of up to 90 dB above threshold at the boundary of the Greek house lot, say 10 m from the rock band's amplifier. Assume that the intensity falls off as the inverse square of the distance *r* from the amplifier. (1) What is the dependence of the dB level on *r*? (2) How far from the amplifier do I have to get before I can converse easily? (Polite speech is about 60 dB above the threshold of hearing.)
(*b*) Fiber-optic waveguides, developed extensively in the Bell Companies, are used to transmit laser light signals in intra-urban communications systems. In 1981, fibers with attenuation losses of 0.2 dB/km were used. If attenuation is exponential with distance, what separation of repeater stations is needed if light intensity ratios between transmitted and received signals of 10^4 are just tolerable? (For more on optical fiber transmission systems, read the book by Personick.)
(*c*) Suppose that college professors lecture at 70 dB above threshold (10^{-12} W/m²), that students' ears are 5 cm² in cross section, that classes meet for 5 hours a week, 30 weeks of the year. If lecturers were paid for their audio output at 6 ¢/kwhr (a 1983 rate for electrical energy), what would their typical annual salary be? (Assume that students listen with both ears.)

[A6.6] {3}   (Reaping Nature's bounty) A problem involving a nonlinear differential equation that can be cast in the form of the logistic growth curve in A6.5 is the effect of harvesting a biological resource such as fish or whales. This problem should therefore be of interest to those concerned with world food supplies or endangered species. Suppose that the logistic growth differential equation (A6.27) is modified to

$$D_t N(t) = \lambda N(t) [1 - N(t)/N_m] - EN(t). \tag{A6.50}$$

The modification consists of a harvesting (or poaching) term, $EN(t)$ in which $E (\geq 0)$ is constant with time. Therefore, we are assuming that the effort to reduce the number of a species $N(t)$ is proportional to their present number. We recall from A6.5 the interpretation of λ as the constant determining the natural rate of increase of the species in the absence of predation and at times when the population is much less than the equilibrium value N_m.
(*a*) Show that with the modified model for population growth the differential equation becomes

$$D_t N(t) = \lambda' N(t)[1 - N(t)/N_m'], \qquad (A6.51)$$

where

$$\lambda' = \lambda - E, \quad N_m' = N_m(1 - E/\lambda). \qquad (A6.52)$$

(*b*) Show that two limits to the behavior of the population for large times, $N(t) \to N_m(1 - E/\lambda)$ if $E < \lambda$, and $N(t) \to 0$ exponentially if $E > \lambda$.
(*c*) Interpret the results in part (*b*) in terms of the harvesting effort per unit population, E, and natural reproduction before self-inhibition is significant, λ.

[A6.7] {3}   (The struggle for survival) Military strategy is, regrettably, an applied science. The English aeronautical engineer, F. W. Lanchester (1868–1946), successfully modeled the attrition of opposing armies in warfare. Consider two opposing armies. Suppose that at time t after the start of a battle there remain $F(t)$ friendlies and $E(t)$ enemies. Lanchester's model assumes that each side has losses in proportion to the size of its enemy.
(*a*) Prove from general sociological principles that any mathematical model for this system must remain unchanged when all labels for E and F are interchanged.
(*b*) Show that for Lanchester's model the appropriate differential equations are

$$D_t F = -\ell_E E, \quad D_t E = -\ell_F F, \qquad (A6.53)$$

where ℓ_E and ℓ_F are the "lethalities." of the enemy and friendly forces. By appropriate application of derivatives show that

$$D_t^2 F = a^2 F, \qquad (A6.54)$$

where $a = \sqrt{(\ell_E \ell_F)}$, with a similar equation for E.
(*c*) Derive solutions for $F(t)$ and $E(t)$ in terms of the hyperbolic functions (section A2.4), namely,

$$F(t) = A_F \cosh(at) + B_F \sinh(at), \qquad (A6.55)$$

$$E(t) = A_E \cosh(at) + B_E \sinh(at), \qquad (A6.56)$$

where A_F, B_F, A_E, and B_E are constants.
(*d*) By using the initial values of F and E, with the original differential equations, show that the constants can be determined, resulting in

$$F(t) = F(0) \cosh(at) - \sqrt{(\ell_E/\ell_F)} \sinh(at), \qquad (A6.57)$$

and, by the symmetry derived in (*a*),

$$E(t) = E(0) \cosh(at) - \sqrt{(\ell_F/\ell_E)} \sinh(at). \qquad (A6.58)$$

(e) Derive Lanchester's law of military survival

$$\ell_F[F^2(0) - E^2(t)] = \ell_E[E^2(0) - E^2(t)]. \qquad (A6.59)$$

(f) Do you think that Lanchester's model is realistic for warfare involving remote control, such as intercontinental ballistic missiles, and mass-destruction weapons, such as nuclear weapons?

(g) For relaxation, or as a displacement activity from militarism, establish whether the board game of checkers (draughts) follows Lanchester's law. To do this, play several games against the same opponent and tally the number of pieces that each of you has on the board after every few moves. Do the number of your pieces (F) and the number of pieces that your opponent has (E) approximately satisfy relation (A6.59)?

For more on Lanchester's differential equation see Dym and Ivey, Chapter 8. Various combat models are compared against data in chapter 8 of the text edited by Braun, Coleman, and Drew. The prediction of wars by computer modeling is discussed in an interesting article by Schrodt in *Byte* magazine. The article also provides a program in Pascal. The statistics of wars from 1816 to 1965 are tabulated in the book by Singer and Small. This sociological problem is also investigated in computing laboratory L8.3.

[A6.8]   {2}   (Unstable methods for differential equations)

(a) Examine both the unstable problem and effects of round-off error by using a computer or calculator program to solve $y'(x) = ay(x)$ using the Simpson rule method, equation (A6.42), for a significant range of a and k values.

(b) Compare your numerical values of $y(x)$ with the "exact" values $y(x) = \exp(ax)$ for the choice of overall scale factor in this linear differential equation, $y(0) = 1$.

(c) Examine the sensitivity to the step-size h by varying it from say 0.05 to 0.5 for each value of the constant a. What is an appropriate constraint relation to make between h and the value of a?

References on differential equations

Acton, F. S., *Numerical Methods That Work*, Harper and Row, New York, 1970.

Alexandrov, I. and P. Lucht., "Physics of Sprinting," *American Journal of Physics*, **49**, 254 (1981).

Blum, H., "Physics and the Art of Kicking and Punching," *American Journal of Physics*, **45**, 61 (1977).

Braun, M., C. S. Coleman, and D. A. Drew, *Differential Equation Models*, Springer-Verlag, New York, 1983.

Cameron, J. R. and J. G. Skofronick, *Medical Physics*, Wiley, New York, 1978.

Clark, C. W., *Mathematical Bioeconomics*, Wiley, New York, 1976.

Dym, C. L. and E. S. Ivey, *Principles of Mathematical Modeling*, Academic Press, New York, 1980.

Feld, M. S., R. E. McNair, and S. R. Wilk, "The Physics of Karate," *Scientific American*, April 1979, p.150.

Gold, H. J., Mathematical Modeling of Biological Systems, Wiley, New York, 1977.

Haberman, R., *Mathematical Models in Mechanical Vibrations, Population Dynamics, and Traffic Flow*, Prentice-Hall, Englewood Cliffs, N.J., 1977.

Hamming, R. W., *Numerical Methods for Scientists and Engineers*, McGraw-Hill, New York, 1962.

Hornbeck, R. W., *Numerical Methods*, Quantum Publishers, New York, 1975.

Keller, J. B., "A Theory of Competitive Running," *Physics Today*, September 1973, p.43.

Lapp, R. E., *The Logarithmic Century*, Prentice-Hall, Englewood Cliffs, N.J., 1973.

McMahon, T. A., and P. R. Greene, "Fast Running Tracks," *Scientific American*, December 1978, p.148.

Meinel, A. B., and M. P. Meinel, *Applied Solar Energy*, Addison-Wesley, Reading, Mass., 1976.

Montroll, E. W. and W. W. Badger, *Introduction to Quantitative Aspects of Social Phenomena*, Gordon and Breach, New York, 1974.

Neie, V. E., "Analysis of Running on Banked and Unbanked Curves," *The Physics Teacher*, **19**, 321 (1981).

Personick, S. D., *Optical Fiber Transmission Systems*, Plenum Press, New York, 1981.

Pizer, S. M. and V. L. Wallace, *To Compute Numerically*, Little, Brown, Boston, 1983.

Riegel, P. S., "Athletic Records and Human Endurance," *American Scientist*, **69**, 285 (1981).

Ryder, H. W., H. J. Carr, and P. Herget, "Future Performance in Footracing," *Scientific American*, June 1976, p.109.

Schrodt, P. A., "Microcomputers in the Study of Politics; Predicting Wars with the Richardson Arms-Race Model," *Byte*, July 1982, p.108.

Singer, J. D., and M. Small, *The Wages of War* 1816–1965, Wiley, New York, 1972.

Thompson, C. D., "Fourier Transforms in VLSI," *International Conference on Circuits and Computers*, IEEE, 1980, p.1046.

Vandergraft, J. S., *Introduction to Numerical Computations*, Academic Press, New York, 1978.

A7 SECOND-ORDER DIFFERENTIAL EQUATIONS

In Chapter A6 you were introduced to the analytic and numeric investigation of differential equations of types frequently encountered in applied science. From the many examples, exercises, and problems in A6 you will have gained an appreciation and a technical knowledge of first-order equations through applications such as the relation between student score and work, the kinematics of footraces, exponential growth and decay, nonlinear equations, and the methods of setting up and solving first-order differential equations numerically.

Why are second-order equations so common?

This chapter emphasizes second-order differential equations. It has much greater emphasis on the dynamics of systems, such as those in mechanics and electricity, than does A6. The reason for this is probably that the fundamental equations of these areas, Newton's and Maxwell's, commonly appear in the form of second-order differential equations.

▶ *Exercise* A7.0.1   Make a list of basic equations in your scientific discipline. Which of them arise from, or appear as, second-order differential equations? Which of them involve first-order equations? ☐

Previewing this chapter, we have in A7.1 an engineering study of hanging cables involving practice in setting up statics problems as differential equations and use of the hyperbolic functions introduced in A2.4. For an interlude, section A7.2 relates the study of catenaries to the history of technology. In A7.3 and A7.4 we return to applicable mathematics in setting up and solving the differential equations for oscillatory mechanical and electrical systems. This leads to resonant behavior, and to the study of resonant line widths. The results of this analysis are applied to atomic clocks in computing laboratory L10. The resting potential across nerve fibers is investigated in A7.5 as an example of a differential equation with applications in biophysics.

Numerical methods for solving differential equations, introduced in A6.6 for first-order equations, are taken up for second-order equations in A7.5. Just as A6.6 emphasized the numerical instability of some first-order equations, A7.7 considers some similarly troublesome second-order equations. Sections A7.6 and A7.7 form the background for the numerical solution of space-vehicle trajectory equations in computing laboratory L11. The inevitable problems and references for further reading conclude the chapter.

130

A7.1 {1}   Cables and hyperbolic functions

The hyperbolic functions were introduced in A2.4 as extensions of the cosine and sine functions to complex angles. In this section they will appear as solutions of differential equations, namely those describing static equilibrium of cables and arches.

   The famous problem of the equilibrium shape of a uniform cable hanging under gravity provides insight into setting up and solving differential equations of second order.

Getting the hang of it

Suppose, as in Figure A7.1.1, that a cable of uniform density hangs in static equilibrium from two points at the same level. Let the weight of the cable per unit length be w. Consider a representative point $P = (x,y)$ and balance the horizontal and vertical force components at P

$$T \cos (\theta) = H, \quad T \sin (\theta) = W = ws, \tag{A7.1}$$

where T is the tension tangential to the cable at P, θ is the angle between the horizontal and the cable, H is the horizontal tension that each side of the cable exerts on the other (it is the same throughout the cable), W is the weight pulling down at the point P. The total weight hanging from P is proportional to the arc length s that lies below P.

   If we combine the formulas in equation (A7.1), we readily derive that the slope of the cable is

$$D_x y = \tan (\theta) = ws/H = s/L, \tag{A7.2}$$

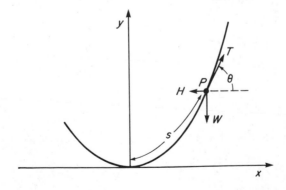

Figure A7.1.1   Forces acting at a representative point P on a uniform cable hanging under gravity.

The parameter L is a length characteristic of the cable and of the local value of gravity. The solution of this differential equation is nontrivial because the arc length depends on x and y.

Interpreting the cable parameter

To get some insight into solving equation (A7.2), let us first consider the interpretation of the cable parameter L.

(1) By writing $wL = H$ we see that L is that length of cable with weight just equal to the tension H applied in a horizontal direction. If you pulled a cable different amounts in the horizontal direction, it would certainly acquire different shapes.

(2) Since $\tan(\theta) = s/L$ has a value of unity at $\theta = 45°$, L is just the length of cable (measured along the curve) from the bottom up to where its slope is $45°$. The tension at this point is $\sqrt{(2wL)}$.

(3) Imagine an extremely heavy chain (such as that for a ship's anchor), the relation $L = H/w$ implies that $L \rightarrow 0$ as w becomes large. So, according to (A7.2), the cable will have a very sharp curve near the bottom. On the other hand, a gossamer spider web arc maintains a very small curvature due to gravity.

▶ *Exercise* A7.1.1

(*a*) To test your insight into the forces involved in a suspended cable, make a rough sketch of how you expect a piece of the same cable hung from two points with a fixed horizontal separation to change shape depending on whether it is on Earth or on the Moon, where gravity is about 1/7 times that on Earth.

(*b*) Describe the shape of the same cable in a space vehicle orbiting Earth. □

Solving the cable differential equation

To obtain the solution of the cable differential equation, we relate the arc length to the slope of the cable by

$$D_x s = \sqrt{[1 + (D_x y)^2]}, \qquad (A7.3)$$

which follows from Pythagoras's theorem applied to a very short section of the cable near P. The arc derivative can also be written in terms of x and y by using (A7.2), so that we have the differential equation for the massy cable

$$D_x^2 y = \sqrt{[1 + (D_x y)^2]}/L. \qquad (A7.4)$$

According to the classification made in section A6.1, this is a second-order nonlinear ordinary differential equation.

To get an idea of the appropriate solution, consider a steep portion of the cable for $x \gg 0$. Here the slope $D_x y \gg 1$, so we can neglect the 1 term on the right-hand side of (A7.4). Now, if we set $p = D_x y$, we find that

$$D_x p \approx p/L. \tag{A7.5}$$

From our exhaustive experience with exponential differential equations in Chapter A6, we recognize immediately that $p \approx c_{1r} \exp(x/L)$, where c_{1r} is a constant, is an appropriate solution for large x. Therefore, integrating this once more, we have

$$y(x) \approx c_{1r} L \exp(x/L) + c_{2r} \tag{A7.6}$$

(where c_{2r} is another constant of integration) is a solution for large x.

▶ *Exercise* A7.1.2   Just to convince yourself that there are no hang-ups in this cable problem, verify all the preceding steps in solving the differential equation for equilibrium. □

In a similar way, we find that for $x \ll 0$, $D_x y$ is negative, and an approximate solution is

$$y(x) \approx c_{1l} L \exp(-x/L) + c_{2l}, \tag{A7.7}$$

where c_{1l} and c_{2l} are also constants of integration.

These solutions for $|x|$ large can be summarized as

$$y(x) \approx L[c_{1l} \exp(-x/L) + c_{1r} \exp(x/L)] + c, \tag{A7.8}$$

where c is an appropriate constant.

From parity (reflection) symmetry in the y-axis, we expect that $c_{1l} = c_{1r} = c_1$ say. Therefore (A7.8) can be rewritten as

$$y(x) \approx 2c_1 L \cosh(x/L) + c. \tag{A7.9}$$

In the last equation we have used the formula for the hyperbolic cosine given in A2.4. How are c_1 and c to be determined?

Another limiting case for the shape of the cable is that for $x \ll L$. Here the cable is relatively flat, so that the derivative term in (A7.4) can be neglected, to obtain for $x \ll L$

$$D_x^2 y(x) \approx 1/L \tag{A7.10}$$

This has the solution by direct integration

$$y(x) \approx x^2/2L + ax + b, \tag{A7.11}$$

where a and b are constants. Again using reflection symmetry about $x = 0$, we require $a = 0$. Let us further define the y origin so that $y = 0$ at $x = 0$. This condition gives $b = 0$. Thus

$$y(x) \approx x^2/2L \qquad (A7.12)$$

for $x \ll L$.

Is this result for small x consistent with that for large x, (A7.9)? If in (A7.9) we require $y = 0$ at $x = 0$, we find $c = -2c_1L$, so the equation for the cable shape becomes

$$y(x) = 2c_1L[\cosh(x/L) - 1]. \qquad (A7.13)$$

Is there a misprint here? Should we have an approximation sign?

▶ *Exercise* A7.1.3   (Proving the cable shape)
(a) Use the Maclaurin expansion of the hyperbolic cosine, (A3.19), to show that for both large and small x this equation is consistent with the approximation (A7.12) obtained for the cable shape, provided that $c_1 = 1/2$.
(b) Verify that the resulting equation for $y(x)$ is an exact solution of the cable differential equation (A7.4). □

Thus, the formula for the height of the cable, y, as a function of the horizontal distance from the center, x, is

$$y(x) = L[\cosh(x/L) - 1]. \qquad (A7.14)$$

Well, this took long enough, but notice that we proceeded from our knowledge of the behavior of actual cables, rather than by mechanically following a canned formula. This approach is more typical of problems in applied science than in pedagogy. Also, I hope that you gained insight into how to arrive at solutions to previously unsolved differential equations. An alternative method of deriving (A7.14), starting from the arc length, is suggested in Problem [A7.1].

Exercises with catenaries

The curve of y versus x described by (A7.14) is called a *catenary*, from the Latin for "chain," as in the verb "concatenate," to chain together. Figure 7.1.2 shows the catenary and its parabolic approximation (A7.12).

▶ *Exercise* A7.1.4   (Catenaries and surveyors)
(a) Show that the length L_c of cable between two points at $\pm x$ is

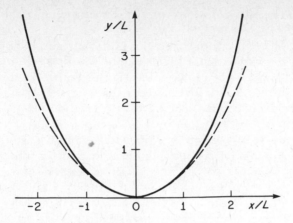

Figure A7.1.2 Catenary function (solid curve) from (A7.14)
and the parabola approximation to it (dashed curve) from
(A7.12).

$L_c = 2L \sinh (x/L)$, where *sinh* is the hyperbolic sine discussed in section A2.4.
(*b*) Show that the tension T at any x is

$$T(x) = H \cosh (x/L). \tag{A7.15}$$

(*c*) A surveyor claims to measure a 50-m horizontal distance to better than 1 mm by
using a 100-m-long steel tape having a mass of 0.5 kg. Considering only the effect
of the weight of the tape, estimate the minimum tension the surveyor must exert on
it to achieve this precision. □

Now that you have worked out some of the analytic properties of the catenary, how about trying a numerical exercise?

▶ *Exercise* A7.1.5   This is a practical exercise in determining L for a cable. The
following data have been obtained for a uniform-density chain with $w = 10$ Nt/m:

x (m)	-3.0	-2.0	-1.0	0	1.0	2.0	3.0	4.0
y (m)	2.7	1.1	0.2	0	0.3	1.0	2.8	5.5

(*a*) Write a pocket calculator or computer program (preferably interactive) that uses
a cut-and-try search method to adjust L to make a least-squares fit (see A4.7) of
these data to the catenary equation (A7.14). Find L to an accuracy of about 5%.
What is a good way to get the search procedure started? Why can't you just use
some canned least-squares formula, such as one of those from A4.7?
(*b*) Make a single graph of the data (as distinct points) and of the fit (as a continuous
curve), appropriately labeled. Also draw the parabolic approximation, $y \approx x^2/L$, for
the full range of x. The plotting procedures in computing laboratory L4 may provide
a satisfactory way of drawing these graphs.
(*c*) How long is the piece of chain between $x = -3$ and $x = 3$? □

A7.2 {1}   Diversion: History of the catenary

The problem of the equilibrium shape of a uniform chain hanging under gravity was posed by Galileo, then by James (Jacob) Bernouilli (1654–1705). It was solved by Robert Hooke (1635–1702), better known for his researches on elasticity and microscopy, in 1676. Proofs that the shape is a catenary were later given by Newton about 1680, by Huygens about 1691, and by Taylor (of Taylor's theorem, A3.3) in 1715.

After the Great Fire of London in 1666, Hooke and Christopher Wren were architectural consultants for the rebuilding of St. Paul's cathedral, and Hooke probably told Wren about the stability of the catenary shape for arches and domes. However, it is unlikely that Wren used this new knowledge when designing the cathedral dome.

The relation between science and technology in the late seventeenth century is discussed in the *Scientific American* article by Dorn and Mark on the architectural work of Christopher Wren. Research on the catenary problem occupied some of the best scientists of the late seventeenth century, but the knowledge does not seem to have been put to much practical use. Two hundred years later, catenary problems were still favorites (of the examiners) in university examinations. The standard text of Routh, *A Treatise on Analytical Statics* (1909), has more than 30 exercises on catenaries. The accelerating pace of science in the late twentieth century is evidenced by the short lead times from scientific discoveries to applications, and by the fact that research problems of the 1940s now appear in college examinations.

▶ *Exercise* A7.2.1
   (*a*) Stone arches in religious buildings often resemble inverted catenaries. (An example is Duke University chapel, a Gothic church built in the 1930s.) What would be a desirable structural property of a catenary arch?
   (*b*) Measure some large archways in churches or other grand buildings. Are they consistent with catenary shapes? What structural constraints would make deviations from the catenary shape either necessary or desirable? □

The interdependence of mathematics, computing, basic science, and technology are emphasized throughout this book. My deliberate intermingling of topics from these areas is intended to persuade you of the concatenations between them and to encourage you to make use of them all when computing in your own scientific discipline.

A7.3 {2}   Second-order linear differential equations

In this section we investigate the differential equations that describe simple mechanical and electrical systems subject to oscillatory motions.

We begin by solving the equations for motion of the system moving without any outside mechanical and electromotive forces. In section A7.4 we use the techniques acquired in the present section to investigate the effects of a driving term on motion within these systems.

Mechanical and electrical analogs

Consider first a mechanical system that models an automobile suspension for a vehicle riding over a bumpy road, as shown schematically in the left diagram in Figure A7.3.1. For simplicity, the bumping forces as a function of time are approximated as having a single frequency. A mass M is subject to forces from a spring with force constant k, from a damper with a force proportional to speed (with proportionality constant b) and to a vertical bumping force from the roadway of $F_0 \cos(\omega t)$. We assume that the weight of the car is balanced at the equilibrium compression of the spring. Motion is assumed to occur only in the vertical direction and to have a value, measured from the equilibrium position, at time t of $y(t)$. Newton's force equation may be used to write

$$D_t^2 y(t) = F_0 \cos(\omega t) - b D_t y(t) - k y(t), \qquad (A7.16)$$

which we write with all the derivatives on one side as the second-order differential equation

$$D_t^2 y(t) + b D_t y(t) + k y(t) = F_0 \cos/\omega t) \qquad (A7.17)$$

This completes setting up the differential equation for the mechanical model.

Consider now the electrical circuit shown in the right diagram in Figure A7.3.1. Under the action of a time-dependent driving voltage of $E_0 \cos(\omega t)$, a time-dependent charge, $Q(t)$, is present in the circuit at time t.

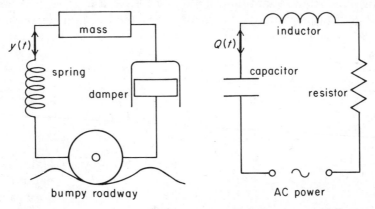

Figure A7.3.1   Mechanical system modeling an automobile on a bumpy roadway (left), and an electrical series circuit with elements having analogous frequency responses (right).

Three circuit components, a capacitor C, an inductor L, and a resistor R, are connected in series. These three components are characterized by the dependence of their impedance (the voltage drop across them per unit current through them) on the frequency, ω, of the driving voltage, as follows: A capacitive element, indicated by C, has an impedance proportional to $1/\omega$. An inductive element, here indicated by L, has an impedance proportional to ω. A resistive element, R, has an impedance independent of frequency. The physical form of these elements is irrelevant, as long as they have these dependences on ω. This is an idealization, since actual inductors, capacitors, and resistors will not have such idealized frequency responses.

To set up a differential equation describing the behavior of this circuit, we use Kirchoff's second law (conservation of energy) to equate the sum of the voltage drops across each circuit element to the source voltage

$$LD_t i(t) + Ri(t) + \int dt' i(t')/C = E_0 \cos(\omega t), \tag{A7.18}$$

where $i(t)$ is the current flowing in the circuit (it's the same everywhere in a series circuit) at time t. This is an awkward integro-differential equation, one that mixes integrals and differentials. However, it can be simplified if we consider the first integral of the current, which is just the total charge in the circuit at time t

$$Q(t) = \int dt' i(t'), \quad D_t Q = i(t), \tag{A7.19}$$

in which the integration extends up to time t. Thus, the differential equation for the charge Q in the circuit at time t becomes

$$LD_t^2 Q(t) + RD_t Q(t) + Q(t)/C = E_0 \cos(\omega t). \tag{A7.20}$$

This equation is very similar to that for the mechanical model, (A7.17). By identifying analogous quantities in the two equations, we can save much effort and gain insight into the behavior of mechanical and electrical systems.

Let us write the general differential equation for forced motion as

$$AD_t^2 y(t) + BD_t y(t) + Gy(t) = S(t) \tag{A7.21}$$

and identify the following analogies:

Mechanical	*Electrical*	*General*	*Interpretation*
M	L	A	inertia
b	R	B	damping
k	$1/C$	G	restoration
y	Q	y	amplitude
$F_0 \cos(\omega t)$	$E_0 \cos(\omega t)$	$S(t)$	source term

To understand how these analogs arise, work the following exercise.

▶ *Exercise* A7.3.1   Consider the preceding table of analogs between mechanical and electrical systems. Give intuitive explanations of these correspondences. To do this, it is probably easiest to discuss the frequency response of each component. For example, do the mechanical inertia (M) and electrical inductance (L) both have a more sluggish response (increasing impedance) as the applied frequency (ω) increases? □

Solving the equations for free motion

We first consider the situation with no driving term in (A7.21); that is, $S(t) = 0$ for all t, postponing the solution of the full equation until section A7.4. The differential equation is then the linear equation

$$A\ AD_t^2 y(t) + BD_t y(t) + Gy(t) = 0. \tag{A7.22}$$

Since springs undulate and circuits oscillate, a solution in terms of cosine and sine functions of t may be appropriate. However, because of damping, we might also expect exponential decay of the amplitude. One way to include both possibilities is to try a solution of the form

$$y(t) = y_0 \exp(mt), \tag{A7.23}$$

in which m is complex, since we know from Chapter A2 that complex exponentials have the desired behavior. With this *ansatz*[1] the derivatives become

$$D_t y(t) = my(t), \ D_t^2 y(t) = m^2 y(t). \tag{A7.24}$$

Substitution into the differential equation (A7.22) yields

$$(Am^2 + Bm + G)y(t) = 0. \tag{A7.25}$$

This has as one solution $y(t) = 0$, for all t. This solution expresses the fact that a system with no external influences [$S(t) = 0$ for all t] will remain undisturbed if it was initially undisturbed.

A more interesting solution of (A7.25) is obtained by solving the quadratic equation for m;

$$m_+ = (-B + \sqrt{D})/2A, \ m_- = (-B - \sqrt{D})/2A \tag{A7.26}$$

---

[1]The German word *ansatz* is sometimes useful if you don't quite know what you're doing, but you need to impress the reader that you do.

in which D is the discriminant

$$D = B^2 - 4AG. \tag{A7.27}$$

Since the differential equation is linear in y, any superposition of two solutions is also a solution. Therefore, the most general solution of (A7.22) is of the form

$$y(t) = y_+ \exp(m_+ t) + y_- \exp(m_- t), \tag{A7.28}$$

where y_+ and y_- are particular values of y determined by the initial conditions. In many applications, particularly when y is the amplitude of a mechanical oscillation or the charge in an electrical circuit, combinations of the basic solutions that make $y(t)$ real are desirable.

Discussion of free-motion solutions

The behavior of y as a function of t depends markedly on the sign of D in (A7.26). We discuss each of the possibilities in turn.

If the discriminant $D > 0$, there are two exponential terms, so that

$$y(t) = y_+ \exp[-(B - \sqrt{D})t/2A] + y_- \exp[-(B + \sqrt{D})t/2A]. \tag{A7.29}$$

This kind of motion is said to be *overdamped*.

▶ *Exercise* A7.3.2   Show that both terms in (A7.29) produce exponential damping, that is $(B - \sqrt{D})/A > 0$, for both the mechanical and electrical systems, unless the mechanical system has a spring with the unlikely behavior that it repels more as the separation is increased. ($k < 0$). [Author's confession: My class notes, as distributed to students, have stated for several years that exponential *increase* is possible. Author's warning: Don't believe everything you read.] □

The typical time dependence for overdamped behavior is shown in Figure A7.3.2.

▶ *Exercise* A7.3.3   Consider the numerical values $B = 5$, $A = G = 2$, with dimensionless variables used for y and t. Show that the motion is overdamped ($D > 0$). With a suitable choice of initial conditions, $y(0) = 1$, $y'(0) = -1$ say, plot $y(t)$ *vs* t, perhaps using the plotting procedure in L4. The curve which you obtain should resemble that for overdamped motion in Figure A7.3.2. □

The second possibility in (A7.26) is that $D = 0$, then the two roots become degenerate, and there is then only a single solution having the exponential form. The solution of the differential equation must be modified to

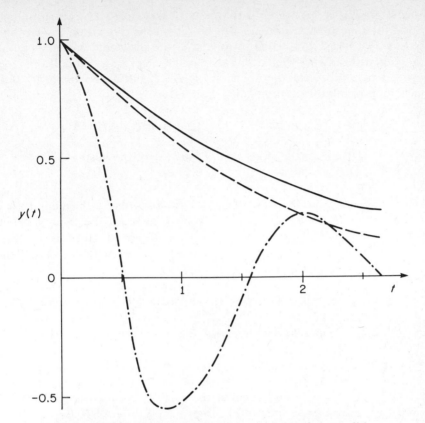

Figure A7.3.2 Overdamped motion (A7.29) shown solid; critically-damped decay (A7.30) shown dashed; oscillatory motion with damping (A7.31) shown dash-dot. Each curve has the same initial value and slope.

$$y(t) = (y_+ + y_- t) \exp(-Bt/2A). \tag{A7.30}$$

▶ *Exercise* A7.3.4   Verify that (A7.30) is indeed a solution of the original differential equation (A7.22), but only in the case that $D = 0$. ☐

For $D = 0$ the motion is said to be *critically damped*. An example of such motion is shown in Figure A7.3.2.

▶ *Exercise* A7.3.5   Consider the example of $B = 4$, $A = G = 2$. Verify that this is critically damped, and plot $y(t)$ versus t for an interesting range of t, using the same initial conditions as in Exercise A7.3.3. The plotting procedure in L4 may be useful. Your curve should be similar to the curve for critical damping shown in Figure A7.3.2. ☐

Critical damping is not merely a mathematical curiosity. A recording instrument used for measuring impulses will be most sensitive when its damping is adjusted to be critical. With this tuning, the rate of decay will be as small as possible, but without the oscillations that make recording difficult.

The third possibility for the solution of (A7.22) attains when $D < 0$. The motion is then oscillatory but steadily damped

$$y(t) = [y_+ \cos (\omega_0 t) + y_- \sin (\omega_0 t)] \exp (-Bt/2A). \tag{A7.31}$$

Here $\omega_0 = \sqrt{|D|}/2A$ is called the *natural frequency* of the system.

▶ *Exercise* A7.3.6
(a) Use the binomial approximation from A3.5 to show that if the damping is very small, then the period of the oscillations $T = 2\pi/\omega$ can be approximated by $T \approx T_0(1 + b^2/8Mk)$, where $T_0 = 2\pi\sqrt{(M/k)}$ is the period if there is no damping.
(b) Give an explanation in terms of physical principles why the period should increase rather than decrease. □

A typical damped oscillatory motion is illustrated in Figure A7.3.2.

▶ *Exercise* A7.3.7   Consider the numerical example (with dimensionless variables) $B = \sqrt{7}, A = 1/2, G = 8$.
(a) Show that $\omega_0 = 3$, and plot the behavior of $y(t)$ *vs* t, using the same initial conditions as in Exercise A7.3.3. The computer plotting procedure from L4 may be a help in this. Your curve should resemble the oscillatory curve in Figure A7.3.2.
(b) Calculate the relaxation time introduced in A6.2, $\tau = 2A/B$. Superimpose the exponential decay factor $\exp (-t/\tau)$ on the graph in (a). □

In this section we have set up and solved a basic second-order differential equation describing a system, mechanical or electrical, that is capable of damped oscillatory motion. In his text Haberman develops these ideas extensively for mechanical vibrations. The next section returns to the problem of forced motion by an external oscillator, particularly emphasizing the nature of resonance phenomena.

A7.4 {2}   Forced motion and resonances

We now return to the problem posed at the beginning of this chapter, motion with a source term, as given by (A7.21). The results to be obtained are very important for studying resonant behavior, with particular application in computing laboratory L10 to the inversion resonance in the microwave spectrum of ammonia. This is analyzed using the Fourier integral transform (see section A5.7) of the Lorentzian resonance line shape to be derived in this section.

Differential equation with a source term

If the source, or driving, term $S(t)$ is nonzero, the method of complex exponentials is useful in solving the differential equation (A7.21). Therefore, in
intermediate steps the variable $y(t)$ will be considered as a complex number.
In most interpretations of $y(t)$ the real part will contain the significant information. We also restrict our considerations to a source term in which there is
a single frequency, ω. (A superposition of frequencies can be handled by the
methods of Fourier expansions developed in A5.) Therefore we write the
source term as

$$S(t) = S_0 \exp(i\omega t). \tag{A7.32}$$

From experience, pumping a swing is a good example, we know that a system under an oscillatory driving influence will settle into steady-state oscillations with the same frequency as the driver. Therefore we try a steady-state
solution of the form

$$y(t) = y_0 \exp(i\omega t), \tag{A7.33}$$

where we expect that the amplitude y_0 will depend on ω. On taking the derivatives of y from (A7.24) with $m = i\omega$ and inserting them in (A7.21), one finds
that

$$[A(i\omega)^2 + B(i\omega) + G]y_0 = S_0. \tag{A7.34}$$

This has the immediate algebraic solution

$$y_0 = -(S_0/A)/(\omega^2 - \omega_0^2 - i\omega\gamma), \tag{A7.35}$$

where the frequency ω_0 is obtained from

$$\omega_0^2 = G/A, \tag{A7.36}$$

and the coefficient of the $i\omega$ term is

$$\gamma = B/A. \tag{A7.37}$$

► *Exercise* A7.4.1   Work through each of the indicated steps in detail to derive
equations (A7.34) through (A7.37) defining the complex amplitude of vibration, y_0.
□

The result (A7.35) for y_0 can be converted into an amplitude and a phase

$$|y_0| = |S_0/A|/\sqrt{[(\omega^2 - \omega_0^2)^2 + \omega^2\gamma^2]}, \tag{A7.38}$$

for the amplitude, and

$$\phi = \arctan[\omega\gamma/(\omega^2 - \omega_0^2)] \tag{A7.39}$$

for the phase angle by which the response $y(t)$ lags the driving term $S(t)$.

Alternative treatment by Fourier transforms

A more general way of deriving these results for response to a driving term, that does not depend on the source term having a single frequency, is to use the Fourier integral transforms developed in A5.7.

We consider the general differential equation for forced motion, (A7.21), multiply both sides by $\exp(i\omega t)$, then integrate over all t. Thus we obtain the Fourier transforms of y, y', and y''.

▶ *Exercise* A7.4.2   Given an integral of the form $\int dt\, D_t^m y(t) \exp(i\omega t)$, integrate by parts m times, and assume that the resultant integrals vanish for very large t, to show that

$$\int dt\, D_t^m.y \exp(i\omega t) = (i\omega)^m Y(\omega), \tag{A7.40}$$

where the integral over t from $-\infty$ to ∞,

$$Y(\omega) = \int dt\, y(t) \exp(i\omega t), \tag{A7.41}$$

is just the Fourier integral transform of $y(t)$. □

From the results of this exercise we have immediately that the response as a function of frequency is just

$$Y(\omega) = -[T(\omega)/A]/(\omega^2 - \omega_0^2 - i\omega\gamma). \tag{A7.42}$$

Here the Fourier transform of the source term, $S(t)$, is $T(\omega)$, where

$$T(\omega) = \int dt\, S(t) \exp(i\omega t), \tag{A7.43}$$

and the other variables are as given in (A7.36) and (A7.37).

The method used here to solve the differential equation with forcing is also that employed in the Fourier transform analysis of resonance line widths in computing laboratory L10.

Resonant oscillations

An important practical question in the design of mechanical structures and of electronic circuits with selective frequency response is "For what driving frequency ω is the frequency response a maximum?'' For maximum response, the denominator in (A7.35) must be a minimum. On differentiating the denominator with respect to ω, one finds that the derivative is zero at $\omega = \omega_R$, where

$$\omega_R(\omega_R{}^2 - \omega_0{}^2 + \gamma^2/2) = 0. \tag{A7.44}$$

The frequency ω_R is called the *resonance frequency* because the response changes most rapidly for ω near ω_R.

▶ *Exercise* A7.4.3   (Resonance frequencies)
(a) Perform the indicated differentiation of (A7.35) to obtain equation (A7.44) for the resonance frequency.
(b) Explain why the solution $\omega_R = 0$ is not acceptable.
(c) Show that there are two values for the resonance frequency

$$\omega_R = \pm\sqrt{(\omega_0{}^2 - \gamma^2/2)}. \tag{A7.45}$$

(d) Explain why only the positive root in equation (A7.45) is usually considered. ☐

Formula (A7.45) for the resonance frequency is usually not as convenient algebraically as an approximation to it, which we now derive.

The Lorentzian function

Here we make some approximations to the resonance formulas, appropriate to the case that $\gamma \ll \omega_0$. By this means we will produce expressions that are both algebraically and numerically tractable, as well as being realistic in most practical applications.

▶ *Exercise* A7.4.4   (Small-damping approximation)
(a) Show that if the damping is relatively small ($\gamma \ll \omega_0$), then

$$\omega_R \approx \omega_0(1 - \gamma^2/4\omega_0{}^2), \tag{A7.46}$$

.that is, the resonance frequency ω_R differs from the natural frequency of vibrations for unforced motions, ω_0, by terms depending on the square of the damping parameter γ.
(b) Assume that ω_0 can be replaced by ω_R in the amplitude formula (A7.38). Thus show that for frequencies close to the resonance frequency the squared amplitude is approximated by

$$|y_0(\omega)|^2 \approx |S_0/(2A\omega_0)|^2 L(\omega), \tag{A7.47}$$

where

$$L(\omega) = 1/[(\omega - \omega_0)^2 + (\gamma/2)^2].\qquad(A7.48)$$

☐

The function $L(\omega)$ is called the *Lorentzian*, after the Dutch theoretical physicist H. A. Lorentz (1853–1928), the godfather of theoretical physics in the early twentieth century. It approximates the frequency dependence of the response of a system as a function of the frequency ω of the driving influence. The general behavior of the Lorentzian is shown in Figure A7.4.1.

Why go to all this trouble to approximate the original frequency-response mula (A7.38) by (A7.47) and (A7.48)? There are two main reasons: First, the differential equation itself is usually only an approximation to the actual system being modeled, so that insistence on a precise solution of it is unrealistic. Second, the Lorentzian function is much easier to handle mathematically than the original form (A7.35). This is illustrated in the following exercise, which derives some of the properties of resonances and Lorentzians.

▶ *Exercise* A7.4.5   (Resonances and Lorentzians)
(*a*) Show that the Lorentzian function is symmetric about ω_0, but that the original response amplitude formula (A7.38) is not.
(*b*) Consider the two frequencies (below and above ω_0) at which the Lorentzian has fallen to half of its maximum value. Show that the separation of these two points is the angular frequency interval, called the *full width at half maximum* (FWHM), γ. ☐

Another way to look at the Lorentzian is from the viewpoint of the complex plane introduced in A2.2. Factor $L(\omega)$ as follows:

$$L(\omega) = [1/(\omega - \omega_0 - i\gamma/2) - 1/(\omega - \omega_0 + i\gamma/2)]/i\gamma.\quad(A7.49)$$

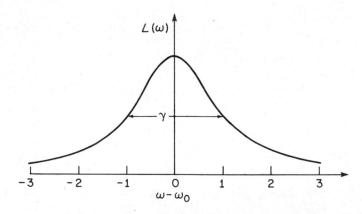

Figure A7.4.1   Lorentzian function $L(\omega)$ as a function of frequency ω and full width at half maximum γ.

If ω were allowed to become complex, then $L(\omega)$ would diverge at $\omega_0 \pm i\gamma/2$. Thus, in the complex angular frequency plane these points are poles of the function $L(\omega)$; only the presence of damping prevents this divergence from occurring in the real world of real frequencies. This extension of a function into the range of complex arguments, so-called *analytic continuation*, was discussed in A2.6 and in Problem [A2.6]. Its application to the Lorentzian function is developed further in Problem [A7.3] (c).

The importance of designing engineering structures so that they do not resonate at frequencies likely to be encountered in use is well illustrated by the occasional collapse from vibration stresses of mechanical structures. Random vibrations can also result in severe structural damage; this topic, and its analysis by Fourier transform methods, are covered in Robson's book. Experiments on the amplitude response of forced oscillations in simple mechanical systems are described in the book by Saraf and coworkers. French's book on vibrations and waves also has a presentation on resonance behavior. Our computing laboratory L10 uses the Lorentzian function in the analysis of data from the inversion resonance of the ammonia molecule, a resonance occurring in the microwave frequency range.

This completes our analysis of simple oscillations in mechanical and electrical systems. Our next example involving a second-order differential equation is a model of conduction in nerve fibers.

A7.5 {2}   Electricity in nerve fibers

When we, or nature, provide an electrically conducting path, the desired path has a much higher conductivity than the surrounding medium. For example, in wired circuits the copper conductor is wrapped in fiber, rubber, or plastic material of low conductivity. However, there is always the possibility of leakage current through this insulating material. How does this modify the voltage distribution along the conductor? We consider this problem for nerve fibers, understanding that the model is also valid (with different parameters) for manufactured electrical conductors such as coaxial cables and power-transmission systems.

The nerve fiber, or *axon*, is a structure within a nerve cell along which electrical signals are conducted. The central core of the axon (the *axoplasm*) is an electrolytic gel, and the membranous wall of the axon, sometimes a sheath of myelin, is not a perfect insulator. We consider only the resting state of axons, since their electrical properties change when they are excited, as is described in the texts by Junge and by Katz.

Modeling nerve fibers

Consider a model of the electrical conduction properties of the axon in which the internal resistance per unit length of the axon is r_i, and the resistance per unit length across the axon to the surrounding cell plasma medium is r_m. The electrical circuit model is shown in Figure A7.5.1. We consider the steady-state voltage as a function of position along the axon x. Call this voltage $V(x)$.

Suppose that along the axon between x and $x + \Delta x$ there is a voltage drop $\Delta V(x)$. In terms of the current flowing along the axon near x, which we call $I(x)$, we have $\Delta V(x) = -I(x)r_i\Delta x$. In the limit $\Delta x \to 0$ we obtain

$$D_x V(x) = -I(x)r_i. \tag{A7.50}$$

We now consider the effect of the leakage current through the walls of the axon to the surrounding medium. This leakage occurs continuously as a function of distance along the axon, although Figure A7.5.1 shows it as occurring at discrete points. Suppose that the leakage current to the cell per unit length is $I_l(x)$, where

$$I_l(x)r_m = V(x). \tag{A7.51}$$

From Kirchoff's law of current conservation applied to a length Δx of the axon near x, the change in current along the axon, $\Delta I(x)$ and the leakage current to the cell are related by $\Delta I + I_l\Delta x = 0$, so that, in the limit $\Delta x \to 0$

$$I_l(x) = -D_x I(x) \tag{A7.52}$$

By combining the last three current and voltage equations, we obtain a second-order differential equation for the voltage $V(x)$ as a function of distance x along the axon

$$D_x{}^2 V(x) = V(x)/\lambda^2, \tag{A7.53}$$

where $\lambda = \sqrt{(r_m/r_i)}$ is constant for a given axon.

▶ *Exercise* A7.5.1   Show in detail the steps leading to the axon-potential differential equation (A7.53). □

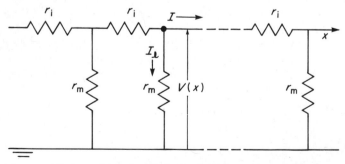

Figure A7.5.1   Electrical circuit model for a nerve cell axon.

Now that we have set up the differential equation for the axon potential, our next task is to solve it.

Solution of the axon potential

You may be nervous about solving this differential equation, but a little action will make it easy. We recall that exponential functions have derivatives proportional to themselves, in the same way that for the axon differential equation (A7.53) the second derivative of the potential is proportional to the potential. Therefore, we try a basic solution of the form $V(x) = \exp(ax)$.

▶ *Exercise* A7.5.2   (Deriving the axon potential)
(*a*) Use the *ansatz* for $V(x)$ in (A7.53) to show that

$$a^2 = \pm 1/\lambda. \tag{A7.54}$$

(*b*) Show that the general solution to the axon equation is therefore

$$V(x) = V_+ \exp(x/\lambda) + V_- \exp(-x/\lambda), \tag{A7.55}$$

where V_+ and V_- depend on the boundary conditions.
(*c*) Assuming that the axon is indefinitely long, derive the final equation for the steady-state potential distribution along a nerve fiber

$$V(x) = V(0) \exp(-x/\lambda). \tag{A7.56}$$

☐

Note that the solution with the increasing exponential would give a shockingly large voltage on a long axon.

As we discussed extensively for exponential behavior in A6.4, the value of λ, the attenuation length, is that length in which the voltage is attenuated by a factor $1/e = 0.368$. The exponential decay of the voltage with distance, and the attenuation length λ, are shown in Figure A7.5.2.

▶ *Exercise* A7.5.3   Suppose that the axon is modeled as a long circular cylinder of radius ϱ, that the specific resistance of the axon per unit volume is R_i and that the resistance of the surrounding membrane per unit area is R_m. Show that the voltage attenuation length becomes

$$\lambda = \sqrt{(\varrho R_m / 2 R_i)}, \tag{A7.57}$$

so that attenuation increases as nerve fiber size decreases, all other things being equal. ☐

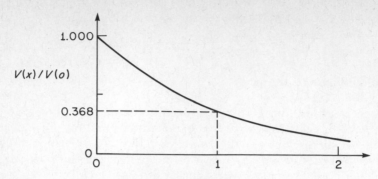

Figure A7.5.2   Potential as a function of distance along a nerve fiber, according to (A7.56). In each attenuation length λ the voltage decreases by a factor 0.368.

Further electrical properties of nerve fibers are explored in Problem [A7.4]. In the text by Junge and in Katz time-dependent axon voltages (nerve signals) are considered, there is a treatment from the medical viewpoint in Cameron and Skofronick, Chapter 3, and the biophysics text by Metcalf has an extensive presentation on nerve cells.

A7.6 {2}   Numerical methods for second-order equations

We return to the investigation of numerical methods for differential equations, begun in section A6.6 with first-order equations. The next two sections emphasize second-order equations, beginning with derivations of Euler algorithms, which are used extensively to compute space-vehicle trajectories in L11. We introduce an alternative algorithm, Numerov's, which is especially efficient for linear second-order equations. Section A7.7 brings up the problems of unstable differential equations (first considered in A6.6) through the problem of the "stiff" differential equation. Various remedies to the difficulties are suggested in the text, exercises, and problems.

Euler approximations

We introduce the Euler approximations for the numerical solution of second-order differential equations in the context of Newton's equations from classical mechanics.

Newton's equations are often solved as a pair of first-order differential equations

$$D_t v = a(x,t), \ D_t x = v(x,t), \tag{A7.58}$$

where v is a velocity component that depends on the acceleration component a in the direction x, say, and on the time t. (Strictly speaking, these are kinematics equations. Newton's equations, which introduce the dynamics, require relating the acceleration to the forces acting.) There are two advantages to this procedure: (1) The techniques for solving first-order differential equations that we developed in A6.6 can be used. This is especially useful for numerical methods. (2) The velocity component v is usually as interesting as x. Note that in our present treatment, unlike in elementary dynamics, the acceleration need not be constant but may depend on position and on time.

The numerical solution of the pair of v and x equations (A7.58) may be developed as follows. Suppose that after n equal time steps each of length Δt the velocity and position are v_n and x_n. By definition, their values after another interval Δt are

$$v_{n+1} = v_n + A\Delta t, \quad x_{n+1} = x_n + V\Delta t, \tag{A7.59}$$

in which A and V are the average acceleration and velocity in this time interval. However, this is not much help because we don't know these values. Therefore, we make various approximations to them.

Our *first Euler approximation* to the averages A and V in (A7.59) is to set

$$v_{n+1} \approx v_n + a_n\Delta t, \quad x_{n+1} \approx x_n + v_n\Delta t, \tag{A7.60}$$

that is, the acceleration and velocity at the beginning of the interval are used to approximate the averages. However, even if the acceleration is constant with time, these approximations do not give the correct formulas.

▶ *Exercise* A7.6.1   Verify the inadequacy of equation (A7.60) by showing that for $a_n = a$, a constant acceleration, the term $a(\Delta t)^2/2$ is missing from the right-hand side of the equation for x_{n+1}. □

To avoid this problem, there are two alternative forms of the Euler approximation. The *second Euler approximation* is

$$v_{n+1} \approx v_n + a_n\Delta t, \quad x_{n+1} \approx x_n + (v_n + v_{n+1})\Delta t/2. \tag{A7.61}$$

Here the average velocity is approximated by the mean of its estimates at the endpoints. (Note that the order of evaluating these two equations is crucial in numerical applications.) The *third Euler approximation* is

$$v_{n+1/2} \approx v_{n-1/2}/a + a_n\Delta t, \quad x_{n+1} \approx x_n + v_{n+1/2}\Delta t, \tag{A7.62}$$

for $n > 0$, and, to get up speed, $v_{1/2} \approx v_0 + a_0\Delta t/2$.

▶ *Exercise* A7.6.2   Show that the second and third Euler approximations (A7.61) and (A7.62) are exact for constant acceleration. □

Our *fourth Euler approximation* for solving the differential equations v and x is

$$v_{n+1} \approx v_n + a_n \Delta t, \quad x_{n+1} \approx x_n + v_{n+1} \Delta t. \tag{A7.63}$$

This uses the acceleration at the beginning of the interval and the velocity at the end as estimators of the mean acceleration and velocity. Although the fourth Euler approximation formulas are not correct for constant acceleration, they turn out to be the best for oscillatory problems, such as harmonic oscillators, and for bound gravitational orbits (as in L11).

The numerical stability of the various Euler approximations applied to solving the motion of a harmonic oscillator is suggested for investigation as Problem [A7.5]. The total energy of a bound, periodic system is conserved if the fourth Euler algorithm is used, as is proved straightforwardly in the article by Cromer. The algorithm seems to have been discovered by a high-school student through a coding error in her program, a serendipity also discussed in Cromer's article.

Numerov's method for linear equations

A very common type of differential equation is the linear second-order equation with the first derivative missing and with y appearing on the right-hand side only as a linear factor, so that we have the linear equation

$$D_x{}^2 y = F(x)y. \tag{A7.64}$$

As a simple example of this type of equation in classical mechanics, we have $x = t$, and $F(t) = -\omega^2$, a constant. The quantity $y(t)$ is then the time-dependent amplitude of vibration of a harmonic oscillator. As an example from quantum mechanics, we have the one-dimensional Schroedinger equation, in which x is the coordinate, $F(x)$ is proportional to the total kinetic energy at x, and $y(x)$ is the probability amplitude at that coordinate.

We develop a very accurate, simple, and efficient algorithm for the numerical solution of equations such as (A7.64). The method is variously attributed to Numerov,[1] or to Fox and Goodwin. It is also discussed in the textbook of Lapidus and Seinfeld on the numerical solution of ordinary differential equations.

We seek a solution to (A7.64) in which only y and its second derivatives appear, and in which an iteration procedure can be used. To simplify the notation we use y''_n to denote $D_x{}^2 y(x_n)$. We seek a two-term iteration relation of the form

$$y_{n+1} + Ah^2 y''_{n+1} = By_n + Ch^2 y''_n + Dy_{n-1} + Eh^2 y''_{n-1} + R, \tag{A7.65}$$

[1] Boris Vasilievich Numerov (1891–1943) was a prominent Russian astronomer and geophysicist.

where A, B, C, D, E are to be chosen optimally, and R is the remainder, which will be used to estimate the error in the algorithm. Consider the Taylor expansion of y_{n+1} about x_n. From the Taylor expansion in A3.3 we have

$$y_{n+1} = y_n + hy'_n + h^2y''_n/2 + \dots, \qquad (A7.66)$$

and the expansion of y_{n-1} is similarly

$$y_{n-1} = y_n - hy'_n + h^2y''_n/2 - \dots. \qquad (A7.67)$$

If we choose $D = -1$ in (A7.65), we can eliminate all the odd-order derivatives. By addition of (A7.66) and (A7.67) we thus obtain the relations

$$y_{n+1} + y_{n-1} = 2(y_n + h^2y''_n/2 + \dots), \qquad (A7.68)$$

and

$$y''_{n+1} + y''_{n-1} = 2(y''_n + h^2y_n^{iv}/2 + \dots), \qquad (A7.69)$$

► *Exercise* A7.6.3 To check whether you are following the details of deriving Numerov's algorithm, pause and derive the preceding results in detail. □

The last equation prompts the choice $E = -A$ in (A7.65) so that the second derivatives occur summed. Equation (A7.65) now becomes

$$y_{n+1} + y_{n-1} + Ah^2(y''_{n+1} + y''_{n-1}) = By_n + Ch^2y''_n + R. \qquad (A7.70)$$

Are we onto a good method? Probably: The formula is symmetric between forward and backward iteration, which should appeal to the aesthetic senses of the mathematician and scientist. Only y and y'', which can be gotten from (A7.64), are involved explicitly, which should appeal to the numerical analyst and computer programmer. So, let us persist in this effort.

Now substitute the Taylor expansions into (A7.70), then equate corresponding derivatives on the left and right sides to show that $B = 2$, $1 + 2A = C$, $A = -1/12$ (and therefore $C = 5/6$). The remainder has leading term $R = -h^6y_n^{vi}/240$.

► *Exercise* A7.6.4 (Details of Numerov's algorithm)
(*a*) Verify the coefficients in the Numerov algorithm by making the indicated substitutions in (A7.70).
(*b*) Carry the Taylor expansions through the sixth derivatives and substitute them into (A7.65) to verify the estimate of the remainder, R. □

The original differential equation, (A7.64), now becomes

$$y_{n+1} - h^2y''_{n+1}/12 = 2y_n + 5h^2y''_n/6 - y_{n-1} + h^2y''_{n-1}/12 + R. \qquad (A7.71)$$

This equation is applicable to any ordinary differential equation. It becomes especially useful when we invoke the linearity condition on the right side of (A7.64), because the second derivatives are then proportional to the function itself. Make the abbreviation

$$Y_n = (1 - h^2 F_n/12)y_n, \tag{A7.72}$$

which appears in (A7.71) when the y'' are substituted by their forms from the original equation. Thus, using the binomial approximation from A3.5,

$$y_n = (1 + h^2 F_n/12)Y_n + \ldots, \tag{A7.73}$$

in which we have made a geometric series expansion of the h^2 factor. Finally, substituting (A7.72) and (A7.73) into (A7.71), we obtain Numerov's algorithm for the solution of (A7.64)

$$Y_{n+1} = 2Y_n - Y_{n-1} + h^2 F_n(1 + h^2 F_n/12)Y_n + R', \tag{A7.74}$$

where the new remainder R' is of order h^6.

▶ *Exercise* A7.6.5   Verify the algebra of the Numerov algorithm by filling in all the steps leading from (A7.71) to (A7.74). □

In summary, Numerov's method for advancing the numerical solution of an equation of the form

$$y''(x) = F(x)y(x) \tag{A7.75}$$

is the following algorithm:

(1) Choose y_0 and y_1 to satisfy the boundary conditions.
(2) Compute Y_0 and Y_1 by using (A7.72).
(3) For $n > 0$ compute

$$D_{n+1} = D_n + h^2 F_n(1 + h^2 F_n/12)Y_n, \tag{A7.76}$$

where $D_0 = 0$, $D_1 = Y_1 - Y_0$. Then

$$Y_{n+1} \approx Y_n + D_{n+1}, \tag{A7.77}$$

with an error of order $h^6 y^{vi}$.
(4) Whenever y_n is needed, use

$$y_n = Y_n/(1 - h^2 F_n/12). \tag{A7.78}$$

This very simple Numerov algorithm for linear second-order equations of the form (A7.75) is much more straightforward to apply than the traditional Runge-Kutta method applied to the same equation.

▶ *Exercise*   A7.6.6   Explain how use of the intermediate variable D_n minimizes round-off errors in the Numerov algorithm. ☐

It is often economical to precompute all the values $h^2 F_n$ if F contains constant factors. If this is done, there are only 7 arithmetic operations needed to advance the solution from n to $n + 1$ in the Numerov algorithm. This algorithm, for linear second-order differential equations of the type (A7.75), is generally much more efficient and easier to program than the Runge-Kutta algorithm. (The Runge-Kutta method, and other numerical methods for ordinary differential equations, are described with Fortran programs in Chapter 8 of the text by Maron.) Moreover, isn't it also more interesting to see some new and improved numerical methods rather than the same old recipes?

Comparison of Euler and Numerov algorithms

One method of comparing the efficiency of two numerical algorithms for solving a differential equation is to compare the total number of arithmetic operations to obtain results of about the same numerical accuracy. For example, if Numerov's method is used with a step size h_N, then there are about 8 operations per step (including the assignment statement), and the error is of order h_N^6. The Euler algorithms need about 6 operations per step, and their error is of order h_E^3, where h_E is the step size used. For results of about the same numerical accuracy in Numerov and Euler algorithms we should therefore choose $h_N \approx \sqrt{h_E}$. However, the number of steps to cover a given range of x is inversely proportional to the step size, so that the larger the h value used, the more efficient is the algorithm. It will also be less sensitive to numerical noise if there are fewer steps.

For example, if an Euler algorithm with $h_E = 0.04$ is sufficiently accurate, then Numerov's algorithm would need only $h_N = \sqrt{0.04} = 0.2$, that is, 5 times fewer steps. would be needed. This far offsets the 8/6 time ratio for each step and is much less sensitive to round-off errors.

We seem to have derived powerful numerical methods for solving differential equations of the form (A7.75), but are they foolproof? The next section shows that some species of these equations can be numerically very troublesome.

A7.7 {3}   Solution of stiff differential equations

The stability of numerical solutions of differential equations was investigated in A6.6 for first-order equations. In this section we extend the discussion to second-order equations, illustrated with (A7.75).

Suppose that in

$$y''(x) = F(x)y(x) \tag{A7.79}$$

we have $F(x) \gg 0$ for a wide range of x, and that solutions of the decaying exponential type are desired. Such a harmless looking differential equation is called a *stiff differential equation*, a terminology that will become clear immediately.

What is a stiff differential equation?

Consider the second-order differential equation (A7.79). For any small range of x values, small enough that $F(x)$ does not change appreciably, the solutions of (A7.79) will be of the form

$$y(x) \approx A_+ \exp[\sqrt{(F(x))}x] + A_- \exp[-\sqrt{(F(x))}x], \qquad (A7.80)$$

as follows from the discussion in A6.4 and may be verified by differentiation of this expression, ignoring the variation of F with x. In most numerical applications the decaying exponential component is the significant one. For example, we may be given the condition $y'(0)/y(0) < 0$. Thus, a numerical solution with $A_+ = 0$ is usually required. If inaccuracies of the numerical integration algorithm and of round-off error allow a small amount of exponentially increasing solution to creep in as x is increased, then this component will quickly grow without bound—stiff luck! This behavior is similar to that discussed in A6.6 regarding the stability of numerical methods for first-order differential equations.

One way to analyze this insidious behavior is to consider what happens if $y(x)$ on the right side of (A7.79) has a small error term ε. The error in $y''(x)$ is then $F(x)\varepsilon$. With $F \gg 0$ this amplifies the curvature of y upward (for $\varepsilon > 0$) or downward ($\varepsilon < 0$), since y''/y indicates the curvature of y when F is large.

▶ *Exercise* A7.7.1   If $F(x) \ll 0$, the solution of (A7.79) remains bounded, but it may be numerically unstable. Justify this statement. Problem [A7.6] shows how to avoid this difficulty. □

One way out of the problem of the stiff differential equation would be to start the numerical integration from the largest x required, using a suitable boundary condition, then to work toward smaller x. Then, although the positive-exponent component of (A7.80) may creep in, it does not grow at the expense of the negative-exponent component. However, it may produce slightly incorrect boundary conditions at small x. If the positive-exponent solution to (A7.79) is required, then the integration should be carried out in the direction of increasing x. These examples illustrate dramatically the noncommutativity of finite-precision arithmetic and approximation formulas.

The Riccati transformation

An alternative solution to the numerical difficulties of stiff differential equations is to transform variables, as follows. The rapidly varying exponential function can be persuaded to softly and suddenly vanish away, like a boojum, by the logarithmic transformation

$$R(x) = D_x \ln(y) = y'(x)/y(x). \tag{A7.81}$$

This is called the *Riccati transformation* of $y(x)$. The motivation for this transformation is shown graphically in Figure A7.7.1. Here the function is $y(x) = \exp(-x)[2 + \cos(x)]$, which graphically resembles a decaying exponential. It decreases by a factor of 60 between $x = 0$ and $x = 3$. However, the value of $y'(x)/y(x)$ oscillates gently, changing by less than a factor of 0.6 over the same range.

▶ *Exercise* A7.7.2   (Properties of Riccati's transformation)
(*a*) Use the Riccati transformation equation in (A7.79) to show that R satisfies the differential equation

$$R'(x) = F(x) - R^2(x). \tag{A7.82}$$

(*b*) Show that if $F(x) \approx a^2$, a constant, then $R(x) \approx \pm a$.
(*c*) Suppose that the differential equation (A7.79) is modified to include a first-derivative term,

$$y'' + by' = F(x)y, \tag{A7.83}$$

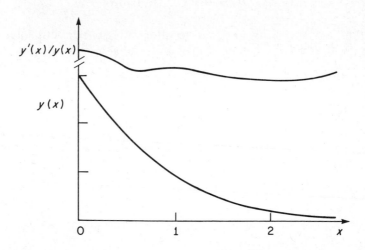

Figure A7.7.1   The Riccati transformation (A7.81) takes an original function with a rapid, approximately exponential, decay (lower curve) and changes it to the well-behaved function shown in the upper curve.

where b is constant. Show that an appropriate transformation is

$$R_b(x) = R(x) + b/2 = y'(x)/y(x), \tag{A7.84}$$

which satisfies

$$R'_b(x) = F(x) - R_b{}^2(x) + b^2/4, \tag{A7.85}$$

which is essentially the same as (A7.82). $\square$

By using these logarithmic transformations we have transformed away the troublesome exponential behavior, at the expense of having to solve a pair of first-order differential equations, first (A7.82) or (A7.85) for $R(x)$, then (A7.81) or (A7.84) for $y(x)$. Care is still needed in solving the latter equations, because exponentially increasing terms can still creep in, but they may be numerically less troublesome than direct solution of (A7.79).

Many of the best computer packages for the solution of differential equations have special procedures for stiff equations. However, these procedures should not be used without caution, because their success or failure will be sensitive to computer-dependent effects, such as the number of significant figures and the round-off algorithms. Practical examples of using the International Mathematical and Statistical Library (IMSL) program package are given in the text by Rice that also describes the PROTRAN system for simplifying Fortran programs that use IMSL procedures.

Problems on second-order differential equations

[A7.1]   {1}   (Cables and arcs) As an alternative method of solving the differential equation for the catenary problem, consider the formula for the arc length

$$D_x s = \sqrt{[1 + (s/L)^2]}, \tag{A7.86}$$

obtained from (A7.2) and (A7.3).

(a) Use the method of separation of variables to show that the arc length is given directly by

$$s(x) = L \sinh (x/L). \tag{A7.87}$$

To do this, you will need to prove the integration formula

$$\int ds'/\sqrt{[1 + (s'/L)^2]} = L \operatorname{arcsinh} (s/L), \tag{A7.88}$$

in which the integration extends from 0 to s.

(*b*) The catenary shape can now be derived using (A7.2) to show that

$$D_x y = \sinh (x/L). \tag{A7.89}$$

Hence show that

$$y(x) = L[\cosh (x/L) - 1] \tag{A7.90}$$

for our choice of coordinates such that $y(0) = 0$.

[A7.2] {2}   (Suspension bridges) As civil engineers (and other polite scientists) know, cables are used in the contruction of suspension bridges. To understand how the cable shape will change depending on the relative weights of vertical members and bridge decking, consider the following examples:
(*a*) Many heavy chains are suspended from a strong cable of relatively negligible weight. The chains have their lowest points at the same level, which is the midpoint of the bridge. Balance forces, as in the catenary example in A7.1, to derive the differential equation

$$D_x y (x) = y(x)/L, \tag{A7.91}$$

where $L = H/w$, using the same notation as in A7.1. Thus show that the cable assumes an exponential shape

$$y(x) = y_0[\exp (|x|/L) - 1]. \tag{A7.92}$$

(*b*) Suppose that a suspension bridge has a deck of weight per unit length H much greater than that of the suspension or vertical cables. Show that the cable shape satisfies the differential equation

$$D_x y(x) = x/L, \tag{A7.93}$$

and therefore that its shape is parabolic,

$$y(x) = x^2/2L, \tag{A7.94}$$

if the coordinate system is chosen so that $y = 0$ at $x = 0$.
(*c*) Plot the catenary (A7.14), exponential, and parabolic functions all on the same graph using as the horizontal scale x/L and as the vertical scale y/L. Choose values of $x/L < 3$. Choose y_0 in (A7.92) so that the exponential and parabolic shapes coincide at $|x| = L$.
(*d*) Explain from force considerations why the three curves in (*c*) are similar for small values of x, yet quite dissimilar for large x values.

[A7.3] {2}   (Properties of Lorentzians) Consider a scaled version of the Lorentzian function (A7.48)

$$L'(\omega') = \gamma'^2/[(\omega' - 1)^2 + \gamma'^2], \tag{A7.95}$$

in which $\omega' = \omega/\omega_0$ and $\gamma' = \gamma/2\omega_0$.

(a) Show that $L'(\omega')$ has a maximum at $\omega' = 1$, and that $L'(1) = 1$ for the maximum value.

(b) Write a program that changes γ' from 0.01 to 1 by factors of 10, and for each value of γ' varies ω' from -1 to 1 by steps of $\gamma'/2$, calculating and plotting $L'(\omega')$ at each point. Thus verify numerically the properties deduced analytically in Exercise A7.4.5. The plotting program in L4 is useful for this assignment.

(c) Investigate analytic continuation of the Lorentzian by calculating $L'(\omega')$ for complex values of ω' with $|\omega'| \leqslant 3$, for a fixed value of γ', say $\gamma' = 0.1$. Make contour plots of the values of L' in the complex ω' plane. Do the divergences (poles) of $L'(\omega')$ occur at $1 \pm i\gamma'$, as the analysis in section A7.4 implies?

[A7.4] {2}   (Electricity in axons) The nerve fiber (axon) was modeled in A7.5 as a lossy conductor. In this problem you will discover other interesting aspects of axon electricity.

(a) To appreciate the biological effectiveness of axons, show that the power drop I^2R_i along a length λ of the axon is insensitive to variations in the axon internal resistance R_i. Thus, temperature and other environmental effects on R_i do not affect the axon performance appreciably.

(b) A crustacean axon has $\varrho = 30$ μm, $R_i = 50$ Ω cm^{-3}, $R_m = 5000$ Ω cm^{-2}. Calculate λ and the energy attenuation in dB/mm. (The decibel unit, dB, is discussed in A6.4.)

(c) Muscle fibers have R_i low compared with nerve fibers, which are usually covered with myelin, which is not a perfect insulator. How does this affect muscle electrical properties compared with those for nerves?

(d) In A7.5 we considered only steady-state conditions. What other properties are important in the transmission of nervous impulses? (For background, see Junge or Katz.)

[A7.5] {3}   (Stability of Euler approximations) The stability of various Euler approximations for the numerical integration of second-order differential equations can be investigated by considering the simple-harmonic oscillator, for which the acceleration $a(x,t) = -\omega^2 x(t)$.

Write and run a computer program and compare the results of $x(t)$ computed using the four different forms of Euler approximations. Also compare these with the values of $x(t)$ calculated from the analytic solution of the differential equation with the same initial conditions. Convenient initial conditions are motion starting from rest with a displacement of unity. Run the calculations for about 5 periods, using a range of time steps Δt from about 0.01 to about 0.1 of the period.

The stability of the fourth Euler algorithm, (A7.63), is discussed in the article by Cromer and is investigated in computing laboratory L11 for the inverse-square gravitational force.

[A7.6] {2} (Madelung transformation for stiff equations) In A7.7 we considered numerical solution of the stiff differential equation (A7.79), in which $F(x) \gg 0$. Suppose that, alternatively,

$$y''(x) = F(x)y(x), \tag{A7.96}$$

with $F(x) \ll 0$ for a wide range of x. We therefore have a differential equation whose solution is highly oscillatory and therefore can be numerically troublesome.

(a) Use the so-called "Madelung transformation" into amplitude r and phase θ, as in A2.2,

$$y(x) = r(x) \exp[i\theta(x)], \tag{A7.97}$$

to show that r and θ must satisfy

$$r''(x) - r(x)\theta'^2(x) = r(x)F_R(x), \tag{A7.98}$$

$$2r'(x)\theta'(x) + r(x)\theta''(x) = r(x)F_I(x), \tag{A7.99}$$

in which the primes denote differentiation with respect to x and $F_R(x)$ and $F_I(x)$ are the real and imaginary parts of $F(x)$.

(b) Show that if F is purely real, then (A7.99) can be integrated to give $\theta'(x) = B/r^2(x)$, where B is a constant of integration. Show also that substitution of this into (A7.98) gives, for real F,

$$r''(x) = r(x)F(x) + B/r^3(x). \tag{A7.100}$$

Thus, if this last equation can be solved, the result can be substituted into the equation for θ', which is a linear differential equation.

(c) Verify that if in (b) $F(x) = -a^2$, then $r(x) = r(0)$, a constant, and $\theta(x) = \pm ax$ are the appropriate solutions of the amplitude and phase differential equations, as expected.

If $F(x)$ consists of regions of x within which $y(x)$ is rapidly varying, separated by regions of slow variation, then a combination of analytical and numerical methods, variously known by the terms, stationary phase, saddle point, steepest descent, and other alpine analogies, may be useful.

References on differential equations

Cameron, J. R., and J. G. Skofronick, *Medical Physics*, Wiley, New York, 1978.

Cromer, A., "Stable Solutions Using the Euler Approximation," *American Journal of Physics*, **49**, 455 (1981).

Dorn, H. and R. Mark, "The Architecture of Christopher Wren," *Scientific American*, July 1981, p.160.

French, A. P., *Vibrations and Waves*, W. W. Norton, New York, 1971.

Haberman, R., *Mathematical Models in Mechanical Vibrations, Population Dynamics, and Traffic Flow*, Prentice-Hall, Englewood Cliffs, N.J., 1977.

Junge, D., *Nerve and Muscle Excitation*, Sinauer Associates, Sunderland, Mass., 1981.

Katz, B., *Nerve, Muscle, and Synapse*, McGraw-Hill, New York, 1966.

Lapidus, L., and J. H. Seinfeld, *Numerical Solution of Ordinary Differential Equations*, Academic Press, New York, 1971.

Maron, M. J., *Numerical Analysis: A Practical Approach*, Macmillan, New York, 1982.

Metcalf, H. J., *Topics in Classical Biophysics*, Prentice-Hall, Englewood Cliffs, N.J., 1980.

Rice, J. R., *Numerical Methods, Software and Analysis: IMSL Reference Edition*, McGraw-Hill, New York, 1983.

Robson, J. D., *Random Vibration*, Edinburgh University Press, 1963.

Saraf, B. (ed.), *Physics Through Experiment*, Vol. 2, *Mechanical Systems*, Vikas Publishing House, New Delhi, 1979.

A8   APPLIED VECTOR DYNAMICS

In previous chapters, especially in A6 and A7, we studied motion in one dimension, usually with x or t as the independent variable. In this chapter we extend our considerations to vector dynamics, in which motion is not restricted to a single dimension. A major purpose of this chapter is to learn more about applying our knowledge of solving differential equations, acquired in A6 and A7, to dynamical systems. In particular, we consider motion in a plane, using plane-polar coordinates. However, this apparent restriction (to two rather than three dimensions) is sufficiently general to enable all problems involving the motion of a body under a central force to be described thereby. As an example, we consider inverse-square forces, with gravitational and electrostatic forces as the ubiquitous examples.

Our aims are to apply the inverse-square-force equations of motion to obtain an analytical solution for planetary orbits and to use numerical methods to calculate trajectories of planets and other satellites. This chapter is also essential preparation for computing laboratory L11 in which we emphasize the numerical details of calculating space-vehicle orbits and trajectories, using the numerical methods for second-order differential equations developed in A7.6.

We begin this chapter with derivations of kinematic equations in plane-polar coordinates, starting from Cartesian coordinates. In section A8.2 central forces, conservation of angular momentum, and Kepler's second law are all discovered. Satellite orbits are the topic of A8.3, in which the elliptical nature of the orbits is proved (Kepler's first law) and the relationship between period and major axis (Kepler's third law) is derived. Section A8.4 is a diversion into the sociology and history of science relating to Kepler's discoveries. A summary of Kepler's laws, and derivation of the inverse-square laws from these three empirical laws, is given in A8.5. Several problems building on ideas developed here are then offered. A variety of interesting readings that extend the material on applied vector dynamics rounds out the chapter.

A8.1 {1}   Kinematics in Cartesian and polar coordinates

We will soon see that for central forces it is sufficient to consider motion in *two* dimensions only. Therefore, two-dimensional polar coordinates, called *plane-polar coordinates* are sufficient for central-force problems. Polar coordinates in *three* dimensions, called *spherical-polar coordinates* are useful for more general problems in which there is also a natural center of symmetry but will not be considered here.

If you are not completely familiar with the relation between plane-polar and Cartesian coordinates, especially the numerical aspects, computing laboratory L2 would be worth your effort. In L2 there are given programs for converting between polar and Cartesian coordinates.

Polar coordinate unit vectors

Suppose that a point P traces out a plane curve, as shown in Figure A8.1.1. In dynamics the description of this curve as a function of time is called a *trajectory*. We may write for the Cartesian coordinates

$$\mathbf{r} = x\mathbf{i} + y\mathbf{j}, \tag{A8.1}$$

in which x and y may change with time, but the unit vectors along the coordinate axes are time-independent. For the polar coordinates

$$\mathbf{r} = r(\cos \theta \mathbf{i} + \sin \theta \mathbf{j}) = r\mathbf{u}_r, \tag{A8.2}$$

in which the second equality defines the radial unit vector at P, $\mathbf{u}_r$. Note that if $\mathbf{r}$ changes with time, then so does this unit vector. Also, if we were foolish (or perhaps, if we were very smart), we might confuse this vector equation with the complex-number description in polar coordinates $z = r(\cos \theta + i \sin \theta)$, as described in A2.2.

Because we are working in two dimensions, there must be two unit vectors. The second one is the tangential unit vector at P, denoted $\mathbf{u}_\theta$. It is perpendicular to $\mathbf{u}_r$ in the plane and rotated $\pi/2$ in the left-hand screw sense from it.

▶ *Exercise* A8.1.1   How could you convey the sense of rotation ''left-hand screw sense'' to an intelligent, nonhumanoid being, for example, a being in an extraterrestrial civilization? To make the problem really challenging, add the restriction that only digital (in the non-anatomical sense) information, not graphics, is to be communicated.

The searches for extraterrestrial intelligence, which must seriously consider questions such as that here, are well described in the books edited by Billingham and by Goldsmith. □

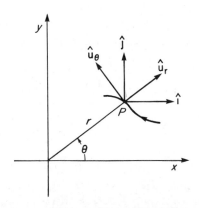

Figure A8.1.1   Motion of a point described in polar and in Cartesian coordinates.

To clarify plane-polar coordinate unit vectors (and to bring you down to Earth) consider the following exercise.

▶ *Exercise* A8.1.2   (Polar-coordinate unit vectors)
(a) Show algebraically that for any motion described by plane-polar coordinates

$$D_\theta \mathbf{u}_r = \mathbf{u}_\theta, \quad D_\theta \mathbf{u}_\theta = -\mathbf{u}_r. \tag{A8.3}$$

(b) Sketch geometrically these two results.
(c) Consider a circular trajectory. Show that the tangential unit vector is always tangential to the circle. □

Velocity and acceleration in polar coordinates

In kinematics we are concerned with describing motion in terms of acceleration and velocity. In polar coordinates, because the two unit vectors depend on each other, motion described in terms of these coordinate systems will sometimes appear more complicated than in Cartesian coordinate descriptions. But, it is only the *description*, and not the *actuality*, that is more complicated. (This is a philosophically biased statement, unpopular since the advent of relativity theory and quantum mechanics.) For example, we speak of centripetal acceleration as a "fictitious acceleration".

To derive the velocity and acceleration relations for plane-polar coordinates in detail we write the displacement as

$$\mathbf{r} = r\mathbf{u}_r. \tag{A8.4}$$

The velocity in plane-polar coordinates can therefore be derived as

$$\mathbf{v} = D_t\mathbf{r} = \mathbf{u}_r D_t r + r D_t \mathbf{u}_r \tag{A8.5}$$

By using (A8.3) in the second term, we have immediately the velocity in plane polar coordinates

$$\mathbf{v} = (D_t r)\mathbf{u}_r + r(D_t \theta)\mathbf{u}_\theta. \tag{A8.6}$$

This is composed of radial and tangential components, respectively. However, as time goes on the directions of these generally change. This is important in what follows.

▶ *Exercise* A8.1.3   Use similar analysis to that in the derivation of the velocity to show that the acceleration in plane polar coordinates is

$$\mathbf{a} = [D_t^2 r - r(D_t \theta)^2]\mathbf{u}_r + [2(D_t r)(D_t \theta) + r D_t^2 \theta]\mathbf{u}_\theta. \tag{A8.7}$$

□

This formula is much more complicated than the equivalent expression in Cartesian coordinates

$$\mathbf{a} = D_t^2 x \mathbf{i} + D_t^2 y \mathbf{j}. \tag{A8.8}$$

To understand the polar-coordinate acceleration formula, some special cases are considered in the next exercise.

▶ *Exercise* A8.1.4   Special cases of formulas (A8.6) and (A8.7) will help you comprehend these complicated formulas.
(a) Suppose that the motion is rectilinear along a fixed angle θ. Show that

$$\mathbf{r} = r\mathbf{u}_r, \quad \mathbf{v} = (D_t r)\mathbf{u}_r,$$
$$\mathbf{a} = (D_t^2 r)\mathbf{u}_r. \tag{A8.9}$$

(b) Consider circular motion, but not necessarily with constant angular velocity. Let the radius of the circle be R. (A good example is a stone being whirled in a vertical plane.) Show that

$$\mathbf{r} = R\mathbf{u}_r, \quad \mathbf{v} = R(D_t\theta)\mathbf{u}_\theta,$$
$$\mathbf{a} = -R(D_t\theta)^2\mathbf{u}_r + R(D_t^2\theta)\mathbf{u}_\theta. \tag{A8.10}$$

(c) Finally, consider uniform circular motion with angular velocity ω. (This can readily be achieved by whirling a stone on a string in a horizontal plane.) Show that

$$\mathbf{r} = R\mathbf{u}_r, \quad \mathbf{v} = R\omega\mathbf{u}_\theta, \quad \mathbf{a} = -R\omega^2\mathbf{u}_r. \tag{A8.11}$$

☐

These exercises show that the formulas for circular motion are just special cases of our polar-coordinate formulas. Several of the terms in the acceleration formula do not have a direct interpretation as being attributable to external forces. It is in this sense that the "centripetal force" is a "fictitious" force. A good question is "Are the common forces, such as gravitation and electromagnetism, just consequences of our choice of coordinate frames?" (If you answer this question definitively, you will probably become more famous than Albert Einstein, who did not find a satisfactory answer to it.)

A8.2 {1}   Central forces and inverse-square forces

In this section we summarize some key results about central and inverse-square forces, especially as they will affect our calculation of analytical and numerical results for calculating satellite orbits and trajectories in computing laboratory L11.

Many of the fundamental interactions describing the forces between two bodies act along the line joining them. We can write this condition in terms of the unit vector along this line as

$$F(r) = F(r)\mathbf{u}_r, \tag{A8.12}$$

in which the function $F(r)$ depends only on the separation r. The direction of the force is specified by $\mathbf{u}_r$. If a force has the property (A8.12), then it is said to be a "central force."

As examples of central forces we have gravitational or electrostatic forces, for which $F(r) \propto 1/r^2$, The force directions are along the vector joining the (point) masses or (point) charges. An example of an interaction for which the force is non-central is that exerted on a charged particle by a magnetic field, for which the fact that the force is perpendicular to the velocity usually does not allow the direction to be along the radius vector, as (A8.12) requires for a central force.

A great simplification for motion of bodies under central forces is reduction of the motion to essentially planar movement, and conservation of angular momentum, as we now derive.

Angular momentum conservation: Kepler's first law

Consider the angular momentum of a body having displacement $\mathbf{r}$ and momentum $\mathbf{p}$ with respect to a fixed point, which is the force center. The angular momentum $\mathbf{L}$ with respect to that point is then

$$\mathbf{L} = \mathbf{r} \times \mathbf{p}, \tag{A8.13}$$

where the $\times$ denotes the vector cross product.

▶ *Exercise* A8.2.1   Calculate the time rate of change of $\mathbf{L}$, and use the fact that the cross product of a vector with itself is zero, to show that $\mathbf{L}$ for a central force does not change with time. ☐

The constancy of the angular momentum about the force center for a central force shows that the motion always lies in the same plane. This plane is identified by locating the force center and any two points of the motion. Alternatively, three independent points of the motion are sufficient to establish the plane of motion. It now becomes clear why our polar coordinates are just *plane*-polar coordinates, rather than full three-dimensional polar coordinates.

▶ *Exercise* A8.2.2   To disprove the converse of the preceding result for the conservation of angular momentum, namely to prove that motion confined to a plane does not necessarily conserve angular momentum, show that rectilinear motion with respect to a fixed point conserves the angular momentum about that point only if the point lies on the trajectory, in which case the angular momentum is zero. ☐

If the central force is gravitational attraction between two bodies, then the constancy of the angular momentum about the force center is called *Kepler's first law*, after Johannes Kepler (1571–1630), a German astronomer who deduced this result from astronomical observations of the planets' motions. Note, however, that only the central nature of gravity is required. Even if the force law were $1/r^3$ (heaven forbid!), the angular momentum would still be constant.

An alternative statement of Kepler's first law is obtained in geometric terms, as follows. Evaluate the magnitude of $\mathbf{L}$ as

$$L = |\mathbf{r} \times \mathbf{p}|, \tag{A8.14}$$

with $\mathbf{p} = m\mathbf{v}$ and the polar-coordinate expression (A8.5) for velocity, to find that

$$L = mr^2 |D_t\theta|. \tag{A8.15}$$

This result leads to the introduction of the concept of areal velocity.

Areal velocity: Kepler's second law

Now let A be the area swept out by the radius vector joining the force center to the other body. This area changes with time at a rate given by, on using (A8.15),

$$D_t A = (D_\theta A)D_t\theta = (r^2/2)D_t\theta = L/(2m). \tag{A8.16}$$

Since L is constant for a central force, we can alternatively say that the "areal velocity", $D_t A$, is constant in time for motion under central forces. This is called *Kepler's second law*. Since Kepler did not have a dynamical model for planetary motions, he stated the two laws as if they were independent; we now know that this is not so. Figure A8.2.1 illustrates the constancy of areal velocity for central forces, showing the particular case of a bound orbit in an inverse-square force.

▶ *Exercise* A8.2.3   (Twists on angular momentum)
(a) Show the detailed steps relating the first two of Kepler's laws for central-force motion.
(b) These results require that the mass m be constant. Would Kepler's laws be modified if relativity theory, instead of Newton's equations of motion, were used?
(c) In the derivations we assumed that the force center was at rest at $r = 0$. One unrealistic way to achieve this is to have an infinitely massive force center (which exerts, unfortunately, an infinite gravitational attraction). What realistic modifications have to be made to these results if the two bodies have comparable masses? (An example of this is a binary star system.) □

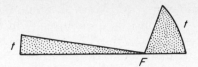

Figure A8.2.1   Constancy of areal velocity about the force
center for any central force. In the same time intervals the
areas swept out by the motion are equal.

As an illustration of the use of Kepler's second law, we can see from (A8.15)
that the angular speed, $D_t\theta$, must increase in inverse-square proportion to the radius r, with a similar decrease if r increases. In particular, in an
unbounded orbit the motion must eventually lie completely along the radius
vector, because it is only thereby that the angular velocity can tend to zero.
This happens in such a way that the area swept out in unit time is constant.

As another example of the use of Kepler's second law for central forces, consider a bounded orbit with turning points at r_p, where the speed is v_p, and at
r_a, where the speed is v_a. At a turning point the velocity must be purely tangential, so that the speed $v = rD_t\theta$. The areal speed is thus given by

$$D_t A = r_p v_p/2 = r_a v_a/2. \tag{A8.17}$$

The angular momentum L is, according to (A8.15), a factor of $2m$ larger than
this.

It is important to know that these results for the constancy of angular
momentum hold for *any* central force, independent of the detailed dependence
of the force on the distance r.

A8.3 {2}   Satellite orbits

In this section we choose a particular central force, the inverse-square force
(as in gravity and electrostatics), and calculate in detail the shape of the orbit
in the case that the motion forms a closed curve, that is, the motion is
bounded.

The *orbit* of a body, such as an earth satellite or solar planet, is a specification of the total path traced out by its center of mass, without reference to the
time at which each point of the motion is reached. In the plane-polar coordinates appropriate to central-force motion in a plane, this amounts to determining $r(\theta)$.

Inverse-square force differential equations

The differential equation for the orbit of a body moving under an inverse-
square central force is tricky to solve, but the method of solution illustrates

how physical intuition about a problem can guide the mathematical formulation and solution, as is demonstrated extensively in A6 and A7.

Consider a satellite of mass m moving outside a much more inert body of mass M. If at some time their separation is r, then (according to Newton) the gravitational attraction between them is

$$F = (- GMm/r^2)\mathbf{u}_r. \tag{A8.18}$$

By using the polar-coordinate expression (A8.10) for the acceleration $\mathbf{a}$, and equating the radial and angular components of the motion that occur on each side of the force equation, we find that

$$D_t^2 r - r(D_t\theta)^2 = - GM/r^2, \tag{A8.19}$$

and

$$2(D_t r)(D_t\theta) + rD_t^2\theta = 0. \tag{A8.20}$$

▶ *Exercise* A8.3.1   Show that the angular equation (A8.20) can be rewritten as

$$D_t(r^2 D_t\theta) = 0, \tag{A8.21}$$

which just verifies Kepler's second law, since gravity is a central force. □

Angular momentum conservation can be used in (A8.19) to eliminate the explicit angular dependence from this equation. As shown in Figure A8.3.1 for the example of an Earth satellite, in terms of radius and speed at perigee

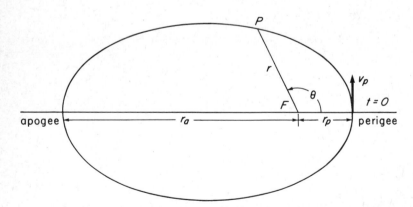

Figure A8.3.1   Coordinate system for analytic solution of orbit equations. For a terrestrial satellite, the distance of closest approach to Earth's center is called perigee and the furthest separation is apogee.

$$D_t\theta = r_p v_p / r^2,\tag{A8.22}$$

so that (A8.19) becomes

$$D_t^2 r = (r_p v_p)^2/3 - GM/r^2.\tag{A8.23}$$

Thus, the net radial acceleration results from the repulsive centripetal acceleration (an artifact of our choice of coordinate system) and the attractive gravitational acceleration (which may be an artifact of our choice of universe). Because of the different dependences of these two forces on r, the motion is quite sensitive to the values of r_p, v_p, G, and M. If these equations are solved numerically, the results are also sensitive to numerical noise, as computing laboratory L11 verifies.

Analytic solution of orbit equations

Our first solution of the equations of motion involves eliminating explicit reference to time, thereby solving for the orbit, the shape of the path traced out in the motion under gravitational forces.

A plane-polar coordinate system and the time origin are chosen as shown in Figure A8.3.1. (The arbitrariness of time origin will be clear to those who have watched the launch of a space vehicle, with occasional holds and clock rollbacks.) Our initial conditions are that the time origin, $t = 0$, is measured from perigee, where $r = r_p$, $D_t r = 0$. For brevity write

$$J = r_p v_p, \quad K = GM,\tag{A8.24}$$

so that the radial acceleration equation (A8.23) becomes

$$D_t^2 r = J^2/r^3 - K/r^2.\tag{A8.25}$$

Since our present interest is in the orbit, we eliminate the independent variable t in terms of θ, as follows:

$$D_t r = (D_\theta r)D_t\theta = (J/r^2) = D_\theta(1/r).\tag{A8.26}$$

By setting the variable $u(\theta) = 1/r(\theta)$ we can rewrite the radial acceleration equation as a differential equation for the orbit

$$D_\theta^2 u(\theta) + u(\theta) = K/J^2.\tag{A8.27}$$

▶ *Exercise* A8.3.2   Transform variables from r to $u(\theta) = 1/r(\theta)$ and evaluate the time derivatives of $u(\theta)$ to derive (A8.27). *Note*: This requires tedious but straightforward applications of the chain rule for derivatives. □

The general solution of the orbit equation (A8.27) can be obtained by inspection

$$u(\theta) = K/J^2 + C \cos \theta + D \sin \theta, \qquad (A8.28)$$

wherein C and D are constants of integration. To see this, note that the first term on the right-hand side is just the solution if u is constant, whereas the circular functions are solutions of (A8.27) when K/J^2 is zero.

▶ *Exercise* A8.3.3   Show that the initial conditions specified in Figure A8.3.1 require that $D = 0$. □

The orbit formula (A8.28) with $D = 0$ can be inverted to give the radius as a function of angle

$$r(\theta) = r_0/(1 + \varepsilon \cos \theta), \qquad (A8.29)$$

where

$$r_0 = J^2/K, \ \varepsilon = JC/K. \qquad (A8.30)$$

If the eccentricity ε is such that $0 \leqslant \varepsilon < 1$, then the orbit describes an ellipse having a semi-major axis

$$a = r_p^2/(2r_p - r_0), \qquad (A8.31)$$

and a semi-minor axis

$$b = a\sqrt{(1 - \varepsilon^2)}. \qquad (A8.32)$$

Since you may not understand as much about conic sections as did the founders of modern mechanics, such as Kepler and Newton, we have a short interlude on the properties of ellipses.

Some geometry of ellipses

The geometry of ellipses is important in visualizing the relations between the dynamics of the satellite and its orbit. This geometry is also important in computing laboratory L11, both in deriving orbit parameters from observables and in displaying space-vehicle orbits.

Ellipses are usually encountered in Cartesian geometry through the equation

$$(x - x_0)^2/a^2 + (y - y_0)^2/b^2 = 1, \qquad (A8.33)$$

where the semi-major axis a and semi-minor axis b are as shown in Fig-

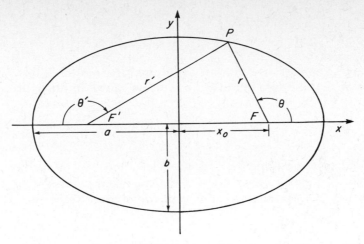

Figure A8.3.2   Geometric quantities for an ellipse with eccentricity 0.8. For inverse-square forces the force center is at one of the foci.

ure A8.3.2, and x_0 and y_0 determine the center of the ellipse relative to the origin of the coordinate system. We derived as (A8.29) that the radius from the force center, r, was related to the angle from perigee, θ, by

$$r(\theta) \;=\; r_0/(1 \,+\, \varepsilon \cos \theta). \tag{A8.34}$$

Our first job is to show that this is consistent with the Cartesian form for the ellipse.

▶ *Exercise* A8.3.4   (Relations for ellipses)
(a) Show by inspection of Figure A8.3.2 and use of the polar equation that

$$a \;=\; r_0/(1 \,-\, \varepsilon^2), \tag{A8.35}$$

and therefore that x_0 in Figure A8.3.2 is given by

$$x_0 \;=\; r_0 \varepsilon/(1 \,-\, \varepsilon^2). \tag{A8.36}$$

(b) Verify that the Cartesian coordinate expression for the ellipse agrees with the polar coordinate form for all θ only if the semi-minor axis b satisfies

$$b \;=\; a\sqrt{(1 \,-\, \varepsilon^2)}. \tag{A8.37}$$

(c) Derive the relation between apogee radius r_a, perigee radius r_p, and the semi-axes a and b:

$$a = (r_a + r_p)/2, \quad b = \sqrt{(r_a r_p)}. \tag{A8.38}$$

☐

For drawing ellipses, as I did when drafting Figure A8.3.2, it is very convenient to know that the sum of the radius vectors shown in the figure is constant.
That is,

$$r(\theta) + r'(\theta') = 2a, \tag{A8.39}$$

for all θ and corresponding θ'. This provides a mechanical method for sketching ellipses; just swing a pencil about the two fixed points F and F' along an
arc formed by a fixed length $(2a)$ of string connected to the two points.

▶ *Exercise* A8.3.5   (Drawing ellipses)
(*a*) Derive equation (A8.39) for the radius vector sum. It may help you to note that
the projections of r and r' onto the y axis are equal, then to eliminate the sine functions in in favor of the cosine functions appearing in the relation (A8.34) between r
and θ. (To change from sine to cosine equate the squares of the projections, then use
the Pythagorean identity $\sin^2\theta = 1 - \cos^2\theta$.)
(*b*) Show that if a circle is viewed from a plane that intersects the x axis such that
there is an angle θ_p between the planes, then the view is that of an ellipse with
$b/a = \cos \theta_p$. This result is important in L11 when choosing eccentricities that can
clearly be seen as elliptical even if the viewing plane is not square-on to the viewer,
as when looking at a book.
(*c*) Suppose that all distances in the y direction are scaled relative to those in the x
direction in the ratio y_s/x_s. Show that the figure of a circle appears as an ellipse. This
result is also important when displaying satellite orbits in L11, because display
devices usually have different distance scales in x and y directions, and therefore will
usually display circles as ellipses unless the different scales are taken into account. ☐

As a remark on changing techniques in science, it is interesting to note that
in Newton's first derivations of the planetary motions he did not use differential calculus and Cartesian geometry. Rather, he used Euclidean geometry all
the way. Nowadays, most of us would be at a loss to solve many problems
without the benefits of Descartes' inventions, Newton's calculus, and plenty
of algebra.

Relation of period to axes: Kepler's third law

You might guess, as Kepler believed and found empirically, that there should
be simple relations between the period of the orbit, T, and the size of the
orbit. We now derive such a relation analytically. From the result for the constancy of the areal velocity,

$$D_t A = r_p v_p/2, \tag{A8.40}$$

we can integrate over one period to obtain

$$A = r_p v_p T/2. \qquad (A8.41)$$

For the elliptical orbits with which we are involved, the area is just $A = \pi ab$. By combining this result with equation (A8.41) for A we get a complicated formula for T. However, this can be simplified to the following remarkable result

$$T^2/a^3 = 4\pi^2/GM, \qquad (A8.42)$$

in which the right-hand side is dependent only on the parent body and not on the satellite.

▶ *Exercise* A8.3.6   Combine the various formulas (A8.24) through (A8.41) to derive equation (A8.42). (I found that the most straightforward way to do this is to write down what T^2/a^3 is according to (A8.41), then to gradually simplify the resulting right-hand side.) □

The relation (A8.42) between the period and the length of the semi-major axis for inverse-square forces is called *Kepler's third law*. What is the interpretation of this result? One way of looking at it is provided by the following exercise.

▶ *Exercise* A8.3.7   Show that Kepler's third law can be expressed as the statement that the period of motion in the elliptical orbit is exactly that of a simple pendulum of length a, oscillating with small amplitude under a gravitational force from a mass M concentrated at the point of suspension. □

Since the gravitational and electrostatic forces have the same distance dependence, the three Kepler laws must also hold for two electric charges moving about each other, except that the force constants are different. You can investigate motion in bound electrostatic orbits in the following exercise.

▶ *Exercise* A8.3.8   An electron moves around a central nucleus of charge Ze. Derive by the analogy between the electrostatic and gravitational interactions the corresponding law for electron orbits moving under electrical forces around atomic nuclei. Note that this result assumes that the motion is governed by the classical equations of motion, Newton's force laws. □

A8.4 {1}   Diversion: Kepler's Harmony of the World

Kepler spent about four years, and 900 pages of hand calculations, reducing data taken by Tycho Brahe between 1575 and 1600 to arrive at a consistent description of the motion of Mars. The procedures that he used are recorded in his *Astronomia Nova,* and they are discussed in the article by Wilson. In 1964, Gingerich repeated Kepler's calculations and used less than 8 seconds on a computer slow by today's standards. He also discovered that Kepler probably made many compensating errors that slowed the completion of his task.

Kepler believed that the motions of the planets about the sun constituted a harmony of the world, and in his book *Harmonices Mundi* he described the varying musical pitches that would be produced by the planets in their orbits if the pitch were proportional to the instantaneous angular velocity. A numerically evaluated, electronic music realization of these harmonies is described in the article by Rodgers and Ruff. Those musically inclined can remember the geometry of Keplerian orbits from the A-major axis to the B-minor axis. Many of Kepler's speculations about the solar system are recorded in his posthumously published work *The Somnium.* For a semi-historical account of Kepler's difficult life read Koestler's, *The Sleepwalkers.* Bernstein has given a semi-popular account of Kepler's passion.

The relation between Kepler's empirical results and the gravitational force was established by Newton. An interesting analysis of Newton's discovery of universal gravitation is given in the *Scientific American* articles by Drake and by Cohen.

By the twentieth century the pace of scientific discovery had accelerated so that the interval between Niels Bohr's contributions to the empirical theory of electron orbits in atoms and the quantum theory was less than 20 years, well within the span of his productive research career.

A8.5 {2}   Summary of Keplerian orbits

Here we bring together Kepler's three laws of bounded motion under inverse-square forces. Then we go on to show that the inverse-square force law follows uniquely from these empirical laws when they are combined with Newton's laws of motion.

Kepler's three laws

We summarize Kepler's laws as he determined them empirically from data on planetary motions, and as Newton derived them in his *Principia*:

Kepler's first law: The bounded orbits of bodies moving under inverse-square forces are ellipses containing the force center in the plane of the motion at one focus of the ellipse.

Kepler's second law: For any central force, the radius vector to the force center sweeps out equal areas in equal times, that is, the angular momentum about the force center is constant.

Kepler's third law: The bounded motion of bodies moving under the same inverse-square-law force center is such that the ratio of the square of the period to the cube of the semi-major axis is the same for all the bodies.

The inverse-square law from Kepler's laws

Suppose, as Kepler found empirically, that these three laws are given. Is there a direct way to prove that the force is *necessarily* an inverse-square force? We proved the converse in the preceding sections. We follow a derivation by Michels, who used energy conservation and some elementary properties of ellipses.

According to Kepler's first law, the orbit is an ellipse. Consider two points P and P' on this ellipse and the radius vectors from them to the foci, as shown in Figure A8.5.1. If a body is at position P, then the velocity $\mathbf{v}$ is perpendicular to the normal OP, and the normal component v_n is perpendicular to the radius vector from the force center F. According to Kepler's second law, the areal velocity is constant, so that

$$v_n r/2 \ = \ \pi ab/T \ = \ b\sqrt{(GM/a)/2}. \tag{A8.43}$$

In the second equality Kepler's third law has been used. In the subsequent steps we use these dynamical results and properties of ellipses to calculate the potential energy as a function of r.

The kinetic energy for a body of mass m is

$$mv^2/2 \ = \ mv_n^2/(2\cos^2 a), \tag{A8.44}$$

where a is shown in Figure A8.5.1. Our task is now to determine a. As you may recall from the optical behavior of an elliptical mirror, the normal OP to the curve at any point bisects the angle between the two radius vectors to the point. We choose the points Q and Q' so that $FQ = FP$ and $F'Q' = F'P'$. Thus, the distances $Q'P$ and QP' are equal, because the sum of the lengths of

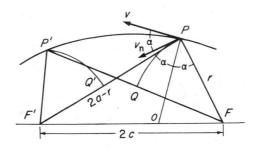

Figure A8.5.1   Geometry of the ellipse for deriving the inverse-square law.

the radius vectors of an ellipse is constant. In the limit of small PQ the figures PQP' and $P'Q'P$ approach congruent right triangles, and the angles become equal to a, as indicated. From the cosine rule in trigonometry applied to the triangle $F'PF$,

$$\cos (2a) = [r^2 - 2ar + 2(a^2 - c^2)]/[r(2a - r)], \qquad (A8.45)$$

so that

$$\cos^2 a = b^2/[r(2a - r)]. \qquad (A8.46)$$

The kinetic energy can now be rewritten from (A8.44) and (A8.45) as

$$mv^2/2 = GM(1/r - 1/2a). \qquad (A8.47)$$

For an isolated system, such as the sun and a single planet, the total energy E is conserved. The potential energy $V(r)$ can therefore be calculated by subtracting the kinetic energy (A8.47) from E. Thus

$$V(r) = E - GMm(1/r - 1/2a). \qquad (A8.48)$$

▶ *Exercise* A8.5.1   Work through each of the trigonometric steps in the preceding derivation of the gravitational potential energy. Note that, like Newton, we require only Euclidean geometry and Kepler's laws, rather than Cartesian geometry, differential calculus and the three laws. □

Since the potential energy in (A8.48) depends only on the distance r, and not on any angles, therefore the force is central. The inverse-square dependence of the force on r follows at once by differentiating (A8.48) with respect to r

$$F(r) = -D_r V(r) = -GMm/r^2. \qquad (A8.49)$$

We note that the quantity GM must be positive, otherwise the kinetic energy would be negative, which is physically unacceptable. Therefore the force is attractive. Note that the existence of a bound elliptical orbit leads to an inverse-square central force. The converse, that inverse-square forces lead necessarily to bound elliptical orbits is contradicted by the phenomenon of nonperiodic comets about the sun. A similar geometric approach to deriving the inverse-square law from Kepler's laws, and a proof that hyperbolic orbits are obtained for repulsive inverse-square forces, has been given by Rainwater and Weinstock.

Problems on vector dynamics

[A8.1]   {2}   (Earth-bound projectiles) In problems about projectile motion under Earth's gravity one uses $D_t^2 r = D_t^2 r = g_k$ and shows that the orbit is parabolic, not elliptical as we derived in A8.3.

(*a*) What simplifications in the polar-coordinate acceleration equations must be made in order that constant acceleration is a good approximation?

(*b*) Estimate the maximum height above Earth and the maximum velocity relative to Earth's surface that a projectile fired at 45° to the local vertical along the equator can have in order that the constant-acceleration approximation is valid to within 1%. Note that all velocities in previous calculations are referred to the force center, but that here it is important to account for the Earth's rotational speed about its axis.

(*c*) Give several reasons why space vehicles launched from Cape Canaveral, Florida, are projected eastward.

(*d*) Kinematically speaking, why is Cape Canaveral a better lunch site than Cape Cod? (Nothing fishy here.)

(*e*) On the Moon, gravity is about one-seventh that at Earth's surface. How does this affect the considerations in part (*b*)?

[A8.2] {3} (Electrostatic-force orbits) Another inverse-square force law to which Kepler's and Newton's results can be applied is Coulomb's force law between electric charges.

As a particular example, consider the orbits of an electron of charge, $-e$ around a hydrogen nucleus (proton) of positive charge e. Assume that Newton's laws of motion hold in the atom, except that (according to Bohr) the angular momentum is quantized according to $L = nh/2\pi$, where n is a non-negative integer and h is Planck's constant, $h = 6.60\times10^{-27}$ erg.sec. The orbits are not necessarily circular. Neglect gravitational forces, but use the analogy between electrostatic and gravitational force laws to carry out the following calculations.

(*a*) Calculate the ratio T^2/a^3 for an electron orbit in hydrogen.

(*b*) What is the period, T, of the $n = 1$ orbit, given that $a = 0.53\times10^{-10}$ m (the Bohr radius) for this orbit.

(*c*) (Especially for chemistry and physics majors) What are the fundamental differences between Bohr's model for electrons in atoms and Newton's model for satellites?

[A8.3] {2} (Solar properties) Did you know that the mass of the sun can be deduced just by knowing some geography and selenography?

(*a*) Show that $GM = g_E r_E^2$, where r_E is Earth's mean radius, 6367 km.

(*b*) Use Kepler's third law, the motion of Earth around the sun, and $G = 6.67\times10^{-11}$ m³/(kg.sec.²), to estimate M_s. The mean distance of Earth from the sun is 1.49×10^8 km, and the orbit may be approximated as circular.

(*c*) The sun and Moon subtend nearly equal angles at a point on Earth, as you will know from the occurrence of nearly exact eclipses. Given that the mean distance between Earth-Moon centers is 3.84×10^5 km, and that Moon's radius is 0.27 that of Earth, estimate the radius of the sun.

(*d*) From the solar mass and radius calculate the mean density of the sun.

Why is not comparable to its planets, such as Earth? (For inhabitants of western seaboards: How can the sun sink into the ocean each evening?)

[A8.4] {2}   (Comets) When a new astronomical object is observed approaching the solar system it is interesting to predict whether it is bound to the sun. Suppose that a comet is first observed when it is at two Earth radii from the sun.

(*a*) If the comet's speed is half that of Earth in its circumsolar orbit, will its orbit be bound?

(*b*) What is your prediction if the comet's speed is twice that of Earth when it is at two earth radii?

(*c*) Halley's comet, with perihelion (closest solar approach) in mid 1986 at 0.587 astronomical units (sun-Earth mean separation), has an eccentricity of 0.967, and a period of 76.2 years. What is its angular velocity about the sun? What is its aphelion (furthest distance)? How much slower is it moving at aphelion relative to its speed at perihelion?

[A8.5] {3}   (Interstellar travel) An interstellar space vehicle in the search for extraterrestrial intelligence (SETI) project is to be launched from Earth.

(*a*) What is its escape velocity from the sun-Earth system? (The mean sun-Earth separation is 1.49×10^8 km.)

(*b*) In terms of energy conservation, there are optimum times and places on Earth from which to launch the SETI probe. Justify your choices for the following optimum conditions for launch, without making a detailed calculation: (1) Optimum time. (2) Optimum position on Earth. (3) Best compass heading. (4) Best elevation above the horizon.

(*c*) Under the optimal conditions from (*b*), what is the minimal launch speed from Earth to escape the solar system?

You can read about the SETI projects in the books edited by Billingham and by Goldsmith.

References on vector dynamics

Bernstein, J., "Kepler: Harmony of the World," in *Experiencing Science*, Basic Books, New York, 1978.

Billingham, J. (Ed.), *Life in the Universe*, MIT Press, Cambridge, Mass., 1981.

Cohen, I. B., "Newton's Discovery of Gravity," *Scientific American*, March 1981, p.66.

Drake, S., "Newton's Apple and Galileo's Dialogue," *Scientific American*, August 1980, p.151.

Gingerich, O., "The Computer Versus Kepler," *American Scientist*, **52**, 218 (1964).

Goldsmith, D. (ed.), *The Quest for Extraterrestrial Life*, University Science Books, Mill Valley, Calif., 1980.

Kepler, J., *Somnium: The Dream, or Posthumous Work on Lunar Astronomy*, University of Wisconsin Press, Madison, 1967.

Koestler, A., *The Sleepwalkers*, Macmillan, New York, 1968.

Michels, W. C., "Rigorous Elementary Derivation of the Inverse-Square Law from Kepler's Laws," *American Journal of Physics*, **41**, 1007 (1973).

Rainwater, J. C. and R. Weinstock, "Inverse-Square Orbits: A Geometric Approach," *American Journal of Physics*, **47**, 223 (1979).

Rodgers, J. and W. Ruff, "Kepler's Harmony of the World: A Realization for the Ear," *American Scientist*, **67**, 286 (1979).

Wilson, C., "How Did Kepler Discover His First Two Laws?," *Scientific American*, March 1972, p.93.

L1 INTRODUCTION TO THE COMPUTING
LABORATORIES

What is the purpose of a laboratory in computing? Some answers to this question, with reference to *Computing in Applied Science*, are given in this introductory chapter. As to how you should use this module, it is important to know that you are not expected to work through each of the L chapters in sequence. Rather, they are to be used as complements to the A chapters.

Analysis and simulation by computer

The speed and reliability of computer operations make the evaluation of complex arithmetic expressions practical in a short time. Several of the computing laboratories emphasize arithmetic computations. For example, laboratory L2 has programs for converting between polar and Cartesian coordinates, L3 emphasizes numerical approximation of derivatives, numerical integration is the subject of L5, and L11 shows how to compute satellite orbits and trajectories.

One consequence of the speed of computers is the possibility for data acquisition and data analysis to proceed in a step-by-step mode. Indeed, many of the techniques developed and practiced in the computing exercises in this L module are suitable for application to such on-line ("real-time") analysis. These data-analysis techniques include spline fitting and interpolation (L7), straight-line least-squares fitting (L8), discrete Fourier transforms, especially by use of the fast Fourier transform (FFT) algorithm (L9), and Fourier integral transforms, as in the analysis of resonance line widths (L10).

Computer graphics, to which we provide an introduction by the widely applicable method of printer plots in L4, can be used whenever it is advantageous for data and analysis results to be displayed. All the computing laboratory exercises from L4 through L11 are greatly enhanced if computer graphics are used to display results.

Simulation, or modeling, of phenomena by computer is often a realistic application of computer capabilities. The applications include design of experiments, as in planning deep-space missions for exploration of the planets, a topic we touch on in the numerical calculation of space-vehicle trajectories in L11.

Random noise in systems can readily be simulated by the Monte-Carlo method, using random numbers. We have extensive exercises on this in L6, including the simulation of round-off errors in computer arithmetic, the approach to equilibrium of a thermodynamic system, and radioactive decay. Noise in signals is also investigated in L9 on the Fourier analysis of EEGs (brain waves) by using a Monte-Carlo simulation.

Coding is not programming is not computing

It is often tempting, and even more frequently practiced, in introductory courses on numerical methods and computer usage to emphasize the detailed instructions in a particular computer language, maybe even for a particular computer in a particular location. This activity of *coding* the computer correctly for a particular machine is a vital part of any automated computation. However, it is often forgotten in the welter of job control language, syntax, and punctuation that the purpose of computing is insight, not numbers.

I assume that you will acquire skill in the use of computer languages from computer programming manuals. This activity I call *programming*. Your skill in programming will certainly be exercised and improved by writing the programs contained in each of the following chapters.

In *Computing in Applied Science* I have emphasized techniques of applicable mathematics and algorithms, rather than specific implementations in a fixed language. This activity I refer to as *computing*. For all except the simplest programs, a program outline in an informal style and using natural language, a "pseudocode," is provided. Only occasionally, when you could probably do with a little boost, or when you might wish to use a general-purpose program (such as the graphics in L4) without having to do all the coding and debugging yourself, do I give you a ready-made and tested program. When I do this, there are two exemplary programs with identical structure, one in Pascal and the other in Fortran. Sometimes I will reference similar programs in BASIC, especially if the algorithm that I present is suitable for this language. The sample programs also give you examples of what I consider fair programming style.

The programming languages

The sample programs in *Computing in Applied Science* are each written in two languages, Pascal and Fortran. To be specific, the Pascal programs were tested on an Apple II + microcomputer running the Apple Pascal language specified in the reference at the end of this chapter. All the Pascal statements that I use are compatible with UCSD Pascal. Therefore, the Pascal programs should run immediately on most computer systems that support Pascal.

The Fortran versions of each program were tested on an Apple II + microcomputer running the Apple Fortran language as described in the reference. The Fortran programs were also tested on a macrocomputer using the WATFIV dialect of Fortran, as described in the book by Dyck, Lawson and Smith. The Fortran statements that I use are compatible with the Fortran 77 standard described in Katzan's book. Most Fortran compilers that support structured-programming statements should be suitable for these programs. To adapt the programs to your computer system you may have to change the input and output statements, especially the way the devices are specified.

Exploring with the computer

If you have a valid model of real-world phenomena programmed in the computer, then, within the limits of the algorithms and numerical methods used, you can explore a wide range of parameter variation within the model. This should lead you to new insights into the phenomena being studied. When the program fails because of limitations of the model or of the computing method, you will learn even more.

A glance at the examples in the L module will show you that most relate to time-dependent effects. For example, all the Monte-Carlo simulations of the real world in L6.3, the least-squares analyses in L8.3, and the space-vehicle trajectories calculated by the numerical solution of differential equations in time in L11.2. This was not deliberate on my part; I think that it can be attributed to human interest in predictive power, from weather forecasting to horoscopes. This power to simulate the future, and therefore to be prepared for or to influence future events, probably constitutes the greatest power of computing in applied science.

References on languages

Apple Fortran Language Reference Manual, Apple Computer Inc., Cupertino, Calif., 1980.

Apple Pascal Reference Manual, Apple Computer Inc., Cupertino, Calif., 1979.

Dyck, V. A., J. D. Lawson, and J. A. Smith, *Introduction to Computing*, Reston Publishing, Reston, Vir., 1979.

Katzan, H., *FORTRAN* 77, Van Nostrand, New York, 1978.

L2 CONVERSION BETWEEN POLAR AND CARTESIAN COORDINATES

The first computing laboratory will give you the opportunity to program straightforward formulas to obtain Cartesian coordinates from plane-polar coordinates. This could be used in converting from the polar-coordinate form $[r,\theta]$ of a complex number, as shown in Figure A2.2.1, for which $z = r \exp(i\theta)$, to the Cartesian form $z = x + iy$. Another application occurs in computing laboratory L11 when transforming the polar coordinates of a satellite position into Cartesian coordinates for the purpose of making a computer graphics display. The conversion to Cartesian coordinates and its programming are described in L2.1.

The inverse transformation, from Cartesian to polar coordinates, is not so straightforward, because it is difficult to identify the quadrant within which θ lies. This more challenging problem and its programming are considered in L2.2.

In both sections of this computing laboratory, sample programs in Pascal and Fortran are given. These programs illustrate basic programming elements in the two languages, so you can also use them for programming practice and review.

L2.1 {1}   Cartesian coordinates from polar coordinates

As in (A2.26) we have the equations for the x and y Cartesian coordinates in terms of the plane-polar coordinates r and θ

$$x = r \cos \theta, \quad y = r \sin \theta. \tag{L2.1}$$

This is straightforward and unambiguous, so it can be expressed in two simple coding instructions. The main points to remember are that r indicates the magnitude of the complex number (or vector in real space), therefore it should always be positive, and that in analytical mathematics θ is usually in radians rather than in degrees.

Pascal program for conversion to Cartesian coordinates

Coding of the Pascal procedures for conversion to Cartesian coordinates from polar coordinates follows (L2.1).

The code presented here executed correctly on an Apple II +, and it uses only Pascal statements compatible with UCSD Pascal.

```
PROGRAM CARPOL;
(* Cartesian from polar coordinates *)
USES TRANSCEND; (* for COS & SIN functions *)
VAR RR,THETA,XX,YY:REAL;
BEGIN
 WRITELN('Cartesian from polar; Pascal');
 WHILE NOT EOF DO
 BEGIN
  WRITELN;WRITE('Input R,THETA:');
  READLN(RR,THETA);
  WRITE('For R=',RR,' & THETA=',THETA);
  XX:=RR*COS(THETA); YY:=RR*SIN(THETA);
  WRITELN(' X=',XX,' Y=',YY);
  END;
END.
```

Exercises using this program are given in Exercise L2.1.1.

Fortran program for conversion to Cartesian coordinates

Equation (L2.1) is used directly to code the Fortran version for conversion to Cartesian coordinates from polar coordinates. The Fortran program has been tested on the Apple II + microcomputer using Apple Fortran, and it has also been tested on a macrocomputer using the WATFIV dialect of Fortran 77.

```
      PROGRAM CARPOL
C  Cartesian from polar coordinates
      REAL RR,THETA,XX,YY
      WRITE(*,10)
  10 FORMAT(' Cartesian from polar; Fortran')
C
  12 WRITE(*,14)
  14 FORMAT(/' Input R,THETA:')
      READ(*,16) RR,THETA
  16 FORMAT(2F5.2)
      WRITE(*,18) RR,THETA
  18 FORMAT(' For R=',F5.2,' & THETA=',F5.2)
      XX = RR*COS(THETA)
      YY = RR*SIN(THETA)
      WRITE(*,20) XX,YY
  20 FORMAT(' X=',F5.2,' Y=',F5.2)
      GO TO 12
      END
```

Exercise L2.1.1 provides the opportunity to use this Fortran program.

Exercises on Cartesian from polar coordinates

The following collection of exercises is divided such that the more parts you work the more elegant (and useful) your own version of the programs just presented will become. However, the later parts of the exercises assume more programming experience.

▶ *Exercise L2.1.1*   (Cartesian coordinates from polar coordinates) The following exercises can be worked using either the Pascal or the Fortran versions of the conversion programs just given.
(*a*) Run the programs for a wide range of θ values (in radians!), and verify by inspection, or by hand calculation, that the correct conversion to Cartesian coordinates, including the quadrant, is obtained.
(*b*) Modify the program you are using so that the condition $r \geqslant 0$ is checked before the conversion to Cartesian coordinates is attempted. If the user has input $r < 0$, a warning message should be output, then your program should input the next set of data.
(*c*) Modify the program CARPOL so that execution is terminated (after a warning message) if $r = 0$ is input.
(*d*) Modify your program so that the user inputs an extra variable, say DEGRAD, which is TRUE if θ is input in degrees, and FALSE if θ is input in radians. In the former case, the appropriate conversion constant from degrees to radians should be applied before the circular functions are evaluated. □

L2.2 {1}   Polar coordinates from Cartesian coordinates

The conversion from Cartesian coordinates to polar coordinates is less direct than the inverse transformation considered in L2.1. Although it is true that

$$r = \sqrt{(x^2 + y^2)}, \tag{L2.2}$$

in which the positive root is taken, the formula

$$\theta = \operatorname{atan}(x/y) \tag{L2.3}$$

is incomplete. One must also examine the signs of x and y in order to determine the quadrant within which θ lies. Therefore, the structured programming statement IF...THEN...ELSE is used frequently in the following programs. These programs provide very good practice in designing and writing programs clearly and concisely.

Figure L2.2.1 shows the Cartesian plane, the corresponding polar coordinates, and the signs of the trigonometric functions in each quadrant. In some computer systems the arctangent function requires two arguments, such as ATAN(X,Y). In this case, the function usually determines the correct quad-

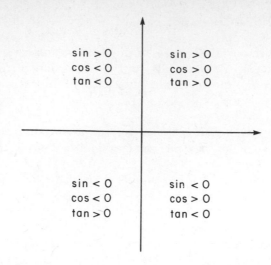

Figure L2.2.1   Signs of the trigonometric functions in each quadrant of the Cartesian plane.

rant. However, if there is only a single argument, as in ATAN(X/Y), there is no way that the correct quadrant can be assigned within the function.

Pascal program for conversion to polar coordinates

The coding of the Pascal version for conversion from Cartesian to polar coordinates is straightforward, provided that care is taken with the choice of the quadrants. The special cases $x = 0$ and $y = 0$ must also be carefully considered.

The Pascal code presented here executed correctly on an Apple II +, using only statements compatible with UCSD Pascal.

```
PROGRAM POLCAR;
(* Polar from Cartesian coordinates *)
USES TRANSCEND; (* for ATAN function *)
VAR PI,XX,YY,RR,ABSTH,THETA:REAL; IQ:INTEGER;
BEGIN
 PI:=4.0*ATAN(1.0); (* PI to machine accuracy *)
 WRITELN('Polar from Cartesian; Pascal');
 WHILE NOT EOF DO
 BEGIN
  WRITELN;WRITE('Input X,Y:');READLN(XX,YY);
  WRITE('For X=',XX,' Y=',YY);
  RR:=SQRT(SQR(XX)+SQR(YY));
  IF RR<>0 THEN
```

```
BEGIN
 IF XX<>0 THEN
  ABSTH:=ATAN(ABS(YY/XX))
 ELSE
  ABSTH:=PI/2.0;
  (* Test quadrants *)
 IF XX>=0 THEN
 BEGIN
  IF YY>=0 THEN
  BEGIN
   THETA:=ABSTH; IQ:=1;
  END
  ELSE
  BEGIN
   THETA:=2.0*PI-ABSTH; IQ:=4;
  END;
 END
 ELSE
 BEGIN
  IF YY>=0 THEN
  BEGIN
   THETA:=PI-ABSTH; IQ:=2;
  END
  ELSE
  BEGIN
   THETA:=PI+ABSTH; IQ:=3;
  END;
 END
 END
 ELSE
 BEGIN
  THETA:=0; IQ:=0; (* Angle undefined at R=0 *)
 END;
 WRITELN(' R=',RR,' THETA=',THETA,' Quad ',IQ);
 END;
END.
```

You can use this program to convert to polar coordinates from Cartesian coordinates in Exercise L2.2.1.

Fortran program for conversion to polar coordinates

I coded the Fortran program directly from the preceding Pascal version, thus ensuring that the logic of determining the correct quadrant for θ was correct. Only a couple of compilations, to correct punctuation and typing errors, were needed to produce a correctly running program. I rechecked computation of angles in all four quadrants and also angles lying along the axes.

The Fortran program was tested on an Apple II + microcomputer using Apple Fortran compatible with the Fortran 77 standard. I also tested it on a macrocomputer (''main-frame'' computer) using the WATFIV dialect of Fortran.

```
      PROGRAM POLCAR
C  Polar from Cartesian coordinates
      REAL PI,XX,YY,RR,ABSTH,THETA
      INTEGER IQ
C   PI to machine accuracy
      PI = 4.0*ATAN(1.0)
      WRITE(*,10)
   10 FORMAT(' Polar from Cartesian; Fortran')
C
   12 WRITE(*,14)
   14 FORMAT(/' Input X,Y:')
      READ(*,16) XX,YY
   16 FORMAT(2F5.2)
      RR = SQRT(XX**2+YY**2)
      IF (RR.NE.0) THEN
        IF (XX.NE.0) THEN
          ABSTH = ATAN(ABS(YY/XX))
        ELSE
          ABSTH = PI/2.0
        ENDIF
C   Test quadrants
        IF (XX.GE.0) THEN
          IF (YY.GE.0) THEN
            THETA = ABSTH
            IQ = 1
          ELSE
            THETA = 2.0*PI-ABSTH
            IQ = 4
          ENDIF
        ELSE
          IF (YY.GE.0) THEN
            THETA = PI-ABSTH
            IQ = 2
          ELSE
            THETA = PI+ABSTH
            IQ = 3
          ENDIF
        ENDIF
      ELSE
```

```
C   Angle undefined at R=0
          THETA = 0
          IQ = 0
       ENDIF
       WRITE(*,18) RR,THETA,IQ
   18 FORMAT(' R=',F5.2,' THETA=',F5.2,
      1 ' Quad ',I1)
       GO TO 12
       END
```

The next subsection will test your skills in coding, programming, and in numerical analysis, that is, your computing skills.

Exercises on polar from Cartesian coordinates

Conversion to polar coordinates from Cartesian coordinates can be checked in the following exercises, which can be worked using either the Pascal or the Fortran implementations of the conversion programs.

▶ *Exercise L2.2.1*   (Polar coordinates from Cartesian coordinates)
(*a*) Run the programs for a wide range of x and y values, to verify that the correct conversion to polar coordinates, especially the correct quadrant, is obtained. Therefore, the relative signs of x and y are more important than their absolute values.
(*b*) Check the consistency of the programs for coordinate conversion by using the output from one as input to the other. Discuss any (small) discrepancies in terms of the accuracy of the cosine, sine, and square root algorithms used in your computer, and in terms of round-off errors (see section A4.2).
(*c*) Modify the program so that if both x and y are zero, then program execution will be terminated after a warning message.
(*d*) Include an extra variable in the program input, called DEGRAD, such that DEGRAD is TRUE if θ is to be output in degrees and FALSE if θ is to be in radians. Check the program for consistency with the output from modification (*d*) in Exercise L2.1.1.
(*e*) If your computer system has an arctangent function with the two arguments x and y, write a small program to compare these arctangent values with the results obtained using your program for the polar angle. How can there possibly be any disagreement? □

The computing practice that you have enjoyed with these coordinate conversion programs will be very useful to you in other computing laboratories, especially those in which the program logic is tricky.

L3   NUMERICAL APPROXIMATION OF DERIVATIVES

The goal of this computing laboratory is to explore techniques for the numerical evaluation of derivatives of functions. We will discover, as section A4.3 reveals in a preliminary way, that numerical differentiation is very sensitive to subtractive cancellation and to other forms of numerical noise, as discussed in A4.2. The hazards and limitations of the techniques will be stressed, particularly because computer evaluation of derivatives is much different from evaluation under direct control and inspection by an experienced numerical analyst.

To explain this remark: In computer calculations we select an algorithm, code and debug a program, and from then on provide input data and accept the program's output. Seldom, unless we suspect severe numerical problems, do we inspect the numbers at many intermediate steps. Therefore we may not be aware of the large changes in the order of magnitude of numbers, of subtractive cancellations, and of truncation problems. In older textbooks on numerical analysis, there are usually chapters on "finite-difference calculus," and many step-by-step exercises to show just what can go wrong with each particular method. It is this step-by-step checking that is ignored in most modern numerical analysis using computers. Furthermore, there is seldom adequate checking built into the programs.

To make you aware of the pitfalls of numerical differentiation, and to give you some methods that may work if judiciously used, this laboratory suggests several exercises applied to generally well-behaved functions. Also, these functions have derivatives that are readily evaluated analytically, thus serving to reveal the inadequacies of the numerical procedures. In particular, we work with functions specified by formulas. This allows us to investigate changing the effects of step size, h. If we try to differentiate data, we are probably limited to the step size at which the data were obtained, so it is usually better to use methods that are not strongly dependent on the choice of h. For data, differentiation is probably best performed by methods of curve fitting such as spline analysis (in A4.6 and L7) and least-squares fitting (in A4.7 and L8).

The background to our computations of numerical derivatives is given in section A4.3. Several of the exercises in that section encourage you to test the numerical methods derived therein. Here we develop a more complete analysis of these topics.

Section L3.1 considers forward and central difference methods and explains Richardson's technique of extrapolation to the limit $h \to 0$. A series of exercises in numerical differentiation is suggested in L3.2. For clarity, we outline a program for investigating numerical derivatives up through the second derivative, and including extrapolation. At the end of the chapter there are references for further reading on numerical derivatives.

L3.1 {1}   Forward and central difference methods

If you have worked the numerical exercises in A4.3, you will have noticed that the forward and central difference methods for numerical derivatives give values that initially improve as the step size, h, decreases, then, on further decreasing h the estimates become very inaccurate. This pathology can be attributed to subtractive cancellation in the numerator of the finite-difference approximation.

One way to avoid the effects of subtractive cancellation is to estimate the derivatives for values of h large enough that subtractive cancellation is not a significant effect, then to extrapolate to infinitesimal h by an algebraic method. We give an example of such extrapolation in the following.

Extrapolation to the limit

We illustrate the method of using finite values of h to infer values for $h \rightarrow 0$ by deriving the formula for central-difference estimates of first derivatives. The general technique is usually called *Richardson extrapolation*. In its use in numerical integration it is referred to as *Romberg integration*. Figure L3.1.1 indicates the extrapolation scheme we use.

From (A4.19) extended to the next nonzero term in h, we have for the central-difference derivative as a function of h

$$D_0(h) = y' + y'''h^2/24 + y^v h^4/1920 + \ldots \qquad (\text{L3.1})$$

For algebraic simplicity we halve the step size at each application of this formula, thereby estimating

$$D_0(h/2) = y' + y'''h^2/24{\times}4 + y^v h^4/1920{\times}16 + \ldots \qquad (\text{L3.2})$$

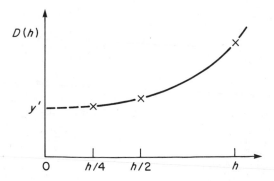

Figure L3.1.1   Extrapolation to the limit for first derivatives.

If we eliminate the third derivative between these two equations, we obtain

$$4D_0(h/2) - D_0(h) = y' - 3y^v h^4/1920 \times 4 + \ldots \tag{L3.3}$$

Replacing h by $h/2$ in this equation, produces

$$4D_0(h/4) - D_0(h/2) = 3y' - 3y^v h^4/1920 \times 4 \times 16 + \ldots \tag{L3.4}$$

Elimination of the fifth derivative from the last two equations, followed by solving for y', results in the final extrapolated equation for the first derivative

$$y' = [64D_0(h/4) - 20D_0(h2) + D_0(h)]/45 + \ldots, \tag{L3.5}$$

in which the error is of order h^6.

Our derivation of the Richardson extrapolation for first derivatives can be generalized, and an algorithm for improving the extrapolation by successive halving of the step size derived, as in the texts by Maron and by Vandergraft. However, our result is sufficient for the following exercises.

L3.2 {1}   Exercises in numerical differentiation

This section outlines a computer program and suggests a set of exercises by which you can develop your skills in numerical analysis as applied to the numerical estimation of derivatives of functions.

Three familiar functions are suggested as test functions: (1) the well behaved circular function $y_1(x) = \sin(x)$, with x in the range 0 to $\pi/2$; (2) the exponential function $y_2(x) = \exp(x)$, which is easily handled for small $|x|$; (3) the hyperbolic function $y_3(x) = \sinh(x)$, which is very well behaved for small x but is as difficult as the exponential for larger argument. There are several advantages to the use of these three test functions. The first is that their derivatives are analytic and simple; (1) $y'_1(x) = \cos(x)$, $y''_1(x) = -y_1(x)$, (2) $y'_2(x) = y_2(x)$, $y''_2(x) = y_2(x)$, (3) $y'_3(x) = \cosh(x)$, $y''_3(x) = y_3(x)$. Another advantage of these functions for testing numerical derivatives is that the circular functions and the exponential function are usually available as part of a computer program library, and the hyperbolic functions *sinh* and *cosh* can be easily formed from the exponential function.

In the following exercises I suggest that you write a basic program, the structure of which is outlined in the next subsection. For each choice of test function replace the function definition and label the output to indicate which function is being used.

Numerical derivatives program structure

The program outline is given here in an informal English pseudocode. It is very straightforward to code in any scientific programming language, such as Pascal, Fortran, or BASIC.

L3 Numerical approximation of derivatives

Definition of function to be differentiated
and its analytic first and second derivatives

Input X, H
First derivatives:
  Forward-difference derivative, D1
  Central-difference derivative at step H, D0H
  Central-difference derivative at step H/2, D0H2
  Central-difference derivative at step H/4, D0H4
  Extrapolate central-difference derivative to get DEH
Second derivative:
  Central-difference derivative at step H, DD0H
Analytic derivatives:
  First derivative, D1; Second derivative, D2

Calculate differences between analytic and numeric derivatives
Calculate Taylor-expansion estimates of errors in each method

Output with labels; X,H,D0H,D0H2,D0H4,DEH,D1,DD0H,D2,
  differences between analytic and numeric,
  Taylor-expansion error estimates

Go for more data

Program end.

With this pseudocode program structure you can code and debug your program for numerical derivatives. How will you debug the program? Well, a linear function for the choice of $y(x)$ should give constant first derivatives in exact agreement with the analytic values and zero second derivatives.

Exercises on numerical derivatives

The following three sets of exercises may be done as you build up the complexity of your program. This is probably a good strategy if you are a novice programmer, or if you are programming in an unfamiliar language.

▶ *Exercise* L3.2.1   (Numerical derivatives of simple functions)
Consider the functions sin (x), exp (x), and sinh (x). Their first and second derivatives are given at the beginning of this section. With each function in turn as the defined function in your program, perform the following calculations:
(*a*) For an interesting variety of x values calculate the forward-difference derivatives $D_+(h)$ for a range of h values in a decade scale. That is, start with say $h = 0.1$, then estimate the derivative using this value, then using $h/10$, $h/100$, until the step size is about the least significant digit that your computer system can handle. You may

save some labor by writing a program loop to do this automatically. Plot $D_+(h)$ against h on a semilogarithmic scale for h. Why do your results initially tend toward a constant value as h decreases, then become erratic as h becomes even smaller?
(b) Repeat this analysis using the central-difference formulas for the first derivative. Are your results in better agreement with analytic results for larger h than are the forward-difference estimates in (a)? If so, why?
(c) For both the forward and central differences, use your plots of the derivative estimate against h to take it to the limit by an eyeball fit. How well do these values agree with the analytic results? □

The next exercise set will allow you to compare your own extrapolation with the Richardson formula (L3.5).

▶ *Exercise L3.2.2*   (Extrapolating to the limit for derivatives)
Consider the functions sin (x), exp (x), sinh (x) for an interesting range of x values. For each function and for each x value, perform the following computations for the first derivative:
(a) Use the Richardson extrapolation formula (L3.5) to take it to the limit.
(b) Compare the extrapolation estimate with the analytic result for this x value and function used.
(c) How well does the algebraic extrapolation agree with your eyeball fit in the previous exercise? Which values do you think are the more reliable? □

Second derivatives are much more challenging than first derivatives because of their greater sensitivity to the step size h. Therefore, in the following exercise it is probably worthwhile to consider only their central difference estimates.

▶ *Exercise L3.2.3*   (Second derivatives numerically estimated)
For each of the functions sin (x), exp (x), sinh (x) use the central-difference second-derivative numerical estimate (A4.22) for a range of x values and step sizes decreasing in a decade scale. For each x do the following:
(a) Compute the second-derivative estimates and their differences from the analytic derivatives for the same argument x.
(b) Compare the actual error in these estimates with that obtained from the first neglected term of the Taylor expansion, as indicated in (A4.22).
(c) What is the dependence of the second-derivative estimates on the value of h and on the number of significant figures carried by the computer you are using? □

Other differentiation techniques

From this computing laboratory you will have learned some of the pitfalls of numerical approximation of derivatives. In particular, the results may be very sensitive to the computer being used because of the large effects that may arise from round-off errors. The Richardson extrapolation technique that you applied here is one way of reducing such errors.

We catalog briefly other techniques of numerical differentiation of functions, or of data tabulated at regular spacings. One technique, much in favor in the days of hand calculations, when error build-up could be inspected (and sometimes controlled), was to fit a polynomial through points near the *x* value at which the derivatives were required. The derivatives were obtained in terms of the analytic derivatives of the polynomial whose coefficients were determined by the fitting. In the handbook of tables and formulas by Abramowitz and Stegun there are many such formulas, and they are also considered in the text on numerical analysis by Hornbeck. A variety of numerical differentiation techniques is provided in the numerical analysis text by Ralston and Rabinowitz.

The method of cubic splines, which we consider in section A4.6 and in computing laboratory L7, is now often considered to be preferable to polynomials, although more numerical work is required. (Just give it to the computer!)

If the function whose slopes are to be found is ill-defined, as for data with significant errors, a least-squares analysis in polynomials may be suitable. We discuss the least-squares principle in A4.7, and computing laboratory L8 is devoted to the fitting of straight lines through such data.

References on numerical derivatives

Abramowitz, M., and I. A. Stegun, *Handbook of Mathematical Functions*, Dover, New York, 1964.

Hornbeck, R. W., *Numerical Methods*, Quantum Publishers, New York, 1975.

Maron, M. J., *Numerical Analysis: A Practical Approach*, Macmillan, New York, 1982.

Ralston, A., and P. Rabinowitz, *A First Course in Numerical Analysis*, 2nd edition, McGraw-Hill, New York, 1978.

Vandergraft, J. S., *Introduction to Numerical Computations*, Academic Press, New York, 1978.

array to which the coordinate (x, y) is transformed be (i_x, i_y). Consider first the How often have you been confronted with a table of 100 pairs of numbers and tried to comprehend them all at once? Haven't you wished that they had also been presented as a graph? As a medical colleague working in computed tomography remarked to me "Computers quantify, people recognize." In applied science the results of computations must often be comprehensible to those who are not expert in computing. Computer graphics provides a convenient way of displaying data that are in computer-accessible form. As with numerical calculations, digital computers provide flexibility, reliability, and repeatability in the generation of plots.

However, very explicit instructions regarding the layout of each plot must be given. Indeed, you may consider it tedious how detailed the instructions must be. But, if you ever tried to give someone else instructions for drafting, or have run off the edge of a sheet of graph paper after casually choosing the scale, then you certainly appreciate why detailed instructions are needed for a brainless computing machine.

The simple computer graphics procedures outlined in this computing laboratory are used extensively in the succeeding laboratories. For example, in L5 equipotential distributions may be plotted by using this procedure, and in L6 the distribution of random numbers may be displayed in a histogram plot modified from this procedure. Most of the other computing laboratory exercises suggest using graphics as a convenient way to visualize and comprehend the results of analyses. If the data are already in the computer, it is most practical for the computer to convert them for graphical representation, rather than to print them out and require the scientist to manipulate them onto graph paper.

Why printer graphics?

The graphics procedures described here use printer output or a video screen functioning as a printer. Although this produces plots with rather coarse resolution and in a single color, it has the merit of being of wide applicability. Other computer graphics methods, especially those with high spatial resolution or multicolors, usually depend strongly on the graphics output device available, on the computer system used, and on the language used to write the graphics procedures and the programs that invoke them. They are briefly considered in L4.3.

The outline of this computing laboratory is as follows: In section L4.1 we introduce plotting using printers, determine how to establish scales and origins, and outline the structure of the printer plotting program. Sample programs for printer plots, in both Pascal and Fortran code, are presented in L4.2, together with plotting exercises. Section L4.3 discusses other graphics techniques, such as the use of dot-matrix printers, vector-graphics devices, and the use of graphics program packages. A set of references on computer graphics rounds out this chapter.

L4.1 {1}   Plotting using printers

In this section you will learn how to make simple, general-purpose plots suitable for output to a printer, a printing terminal, or a video screen with character output. The graphs produced by this method have a resolution limited by that of the printer, typically 132 or 80 characters across the display and 66 or 24 lines in depth. The advantages of the method are that it is very general in applicability, being suitable for any output device that uses characters. The graph may also be stored on permanent storage, such as magnetic disk or tape.

One common way to make a printer plot is to generate one line of the plot at a time, output this line, then go on to the next line. This has the disadvantage that the variables must be sorted to determine their order of output, since most printers cannot be backspaced. This is especially cumbersome when several figures are to be plotted on the same graph.

An alternative method, used here, is to make a two-dimensional character array, the elements of which contain the character to be plotted. We can fill this display array in any order, which removes the requirement of having to sort the variables. When the array is filled it is printed as a block. The technical term for the smallest element of a graphic display is *pixel*, from *pic*ture *ele*ment. For a display of 80 by 24 pixels, less than 2 Kbyte of storage is required, if one character is stored in one byte. For a 132 by 66 display, the storage required is less than 9 Kbyte. Both these storage requirements are modest compared with the storage capacity of even a microprocessor-based system.

Our displays will use the x direction across the page and the y direction up the page. This restriction mainly affects the order of output, so that it is easily modified. However, the coordinate that is plotted along the direction of the paper motion (on a printer) or of the scrolling (on a video terminal) must be output in reversed order, largest values first. (If you ever drafted using India ink you probably used this order to prevent smudging.) In the sample programs that follow, I chose to generate the display array with the y coordinates reversed, then to output the array by increasing array element order in both x and y indexes.

In the following subsections we derive formulas for calculating scales and origins, then outline the structure of the printer plotting procedure in pseudocode.

Formulas for scales and origins

The procedure for choosing scales and origins of graphs can be automated and derived from a formula, which may be surprising to those whose graphs never fit appropriately on the paper they have chosen. We assume here that the graph is to fill all the space in the display array. This is done for simplicity; it is not the best drafting practice, but you can easily modify it to your particular style. We also assume that the display is linear in x and y directions. If not, as in a semilogarithmic display, convert the coordinates first, then apply the procedures to be outlined.

Scales and origins are determined as follows. Let the location in the display array to which the coordinate (x,y) is transformed be (i_x, i_y). Consider first the x coordinate. We write

$$i_x = x_s x + x_0, \qquad \text{(L4.1)}$$

where the scale factor x_s and the origin x_0 are to be determined. Suppose that the minimum x value is x_- and the maximum x value is x_+. We want i_x to range from 1 to N_x, the dimension of the display array for the x direction. On inserting the boundary values in equation (L4.1), you can readily derive that

$$x_s = (N_x - 1)/(x_+ - x_-), \qquad \text{(L4.2)}$$

and that

$$x_0 = 1 - x_s x_- \qquad \text{(L4.3)}$$

For the y direction the boundary values are reversed to allow for the usual direction of motion in a printer or video terminal. Therefore

$$i_y = y_s y + y_0, \qquad \text{(L4.4)}$$

where $i_y = N_y$ for $y = y_-$ and $i_y = 1$ for $y = y_+$. Here N_y is the dimension of the display array corresponding to the y direction. Thus, the mapping for the y coordinates has the y scale factor

$$y_s = (1 - N_y)/(y_+ - y_-), \qquad \text{(L4.5)}$$

and the y origin

$$y_0 = 1 - y_s y_+. \qquad \text{(L4.6)}$$

▶ *Exercise* L4.1.1   Derive the preceding scale-factor and origin formulas, starting from the defining equations. ☐

Note that the scale factor and origin calculations require the extreme values of the x and y coordinates. These are determined straightforwardly by searching through the x and y values separately, as detailed in the MINMAX procedures in the sample programs in L4.2. Such a general-purpose routine is probably best coded as a separate program segment from the plotting routine.

We now have all the ingredients for generating the display array, and we are ready to outline the plotting procedure.

Printer plotting procedure structure

The plotting procedure can be outlined in pseudocode, using the results from the previous subsection.

L4 Printer plotting procedure

Input values:
 NXGRID, NYGRID display array dimensions
 NDATA number of (X,Y) pairs to be plotted
 XLIST, YLIST linear arrays of X and Y values
 XMIN, XMAX, YMIN, YMAX extreme values of X and Y

Compute ratio of display scales, XYFACT
Scale factor and origin in X direction by (L4.2) and (L4.3)
Scale factor and origin in Y direction by (L4.5) and (L4.6)

Make edges on graph in X and Y directions
Initialize rest of graph to ground character
 Loop over IX
  Loop over IY

Locate display points in graph
 Loop over data points I,
 computing X and Y indexes by (L4.1) and (L4.4),
 and locate figure character in graph array
 (Y index is scaled by XYFACT)

Output graph
 Loop over X and Y indexes

Procedure end.

In the next section we convert this pseudocode into procedures in the Pascal and Fortran languages, then there are several exercises using these plotting procedures.

L4.2 {1} Sample programs for printer plots

In this section we present programs for computer generation of plots output to a printer, or video screen in text-output mode. You may be surprised that I have provided working programs in two widely used languages. However, these plotting programs are so useful in the other computing laboratories that I want to encourage you to use them.

The Pascal and Fortran programs in the next subsections have been coded so that they are almost identical. I made some compromises of the elegance of Pascal to achieve this. If you need a program in BASIC, you can either work from the preceding pseudocode, using the complete programs for coding suggestions, or you can copy out the program in the textbook by Bennett.

The data to be plotted are contained in the linear data arrays XLIST and YLIST. The programs assume that these arrays are generated outside the plotting procedure. Often they will be obtained from experimental data, or generated as part of the computer program. For test purposes, as indicated in the sample checking programs, it is most practical (for a few data points) to input the XLIST and YLIST arrays as data to the program. The data that I chose for Figure L4.2.1 are the cosine function over about two-thirds of a period.

Printer plots from Pascal

The coding of the Pascal procedures for printer plots follows in a straightforward way from the pseudocode outline in section L4.1. There are three procedures: PLTCHK is the main procedure used to assign global variables and values, to input the data to be displayed, and to invoke the other procedures. MINMAX procedure searches the linear arrays XLIST and YLIST to find their minimum and maximum values. PLOTP procedure is the Pascal version of the program outlined in the pseudocode for the display.

The code presented here executes correctly on an Apple II +, and uses only Pascal statements compatible with UCSD Pascal.

```
PROGRAM PLTCHK;
(* Printer plots from Pascal; checking program *)
CONST NXGRID=50; NYGRID=22; MXDATA=100;
VAR NDATA,I: INTEGER;
    XLIST: ARRAY [1..MXDATA] OF REAL;
    YLIST: ARRAY [1..MXDATA] OF REAL;
    XMIN,XMAX,YMIN,YMAX,XYFACT: REAL;
PROCEDURE MINMAX;
(* Find limits in first NDATA elements of XLIST *)
VAR IXY: INTEGER;
BEGIN
 XMIN:=XLIST[1]; XMAX:=XMIN;
 YMIN:=YLIST[1]; YMAX:=YMIN;
 FOR IXY:=2 TO NDATA DO
 BEGIN
```

```
    IF XMIN > XLIST[IXY] THEN XMIN:=XLIST[IXY];
    IF XMAX < XLIST[IXY] THEN XMAX:=XLIST[IXY];
    IF YMIN > YLIST[IXY] THEN YMIN:=YLIST[IXY];
    IF YMAX < YLIST[IXY] THEN YMAX:=YLIST[IXY]
   END
END; (* MINMAX *)
PROCEDURE PLOTP;
(* Printer plot of YLIST against XLIST *)
CONST FIGURE='*';GROUND=' ';XEDGE='.';YEDGE=':';
VAR GRAPH:ARRAY[1..NXGRID,1..NYGRID] OF CHAR;
   IX,IY,I:INTEGER;
   XSCALE,XORG,YSCALE,YORG:REAL;
BEGIN
 (* X scale factor, then X origin *)
 XYFACT:=NYGRID/NXGRID; (* ratio of scales *)
 XSCALE:=(NXGRID-)/(XMAX-XMIN);
 XORG:=1-XSCALE*XMIN;
 (* Y scale factor, Y origin *)
 YSCALE:=(1-NXGRID)/(YMAX-YMIN);
 YORG:=1-YSCALE*YMAX;
 (* Make edges on GRAPH *)
 FOR IX:=1 TO NXGRID DO
 BEGIN
  GRAPH[IX,1]:=XEDGE; GRAPH[IX,NYGRID]:=XEDGE;
 END;
 FOR IY:=1 TO NYGRID DO
 BEGIN
  GRAPH[1,IY]:=YEDGE; GRAPH[NXGRID,IY]:=YEDGE;
 END;
 (* Initialize rest of GRAPH to GROUND *)
 FOR IX:=2 TO (NXGRID-1) DO
 BEGIN
  FOR IY:=2 TO (NYGRID-1) DO
  BEGIN
   GRAPH[IX,IY]:=GROUND
  END;
 END;
 (* Locate display points, FIGURE, in GRAPH *)
 FOR I:=1 TO NDATA DO
 BEGIN
  IX:=ROUND(XSCALE*XLIST[I]+XORG); (* X *)
  (* Invert Y *)
  IY:=ROUND(XYFACT*(YSCALE*YLIST[I]+YORG));
  IF IY = 0 THEN IY:=1;
  GRAPH[IX,IY]:=FIGURE
 END;
```

```
(* Output; loop over Y and X indexes *)
FOR IY:=1 TO NYGRID DO
BEGIN
  FOR IX:=1 TO NXGRID DO WRITE(GRAPH[IX,IY]);
  WRITELN
END;
END; (* PLOTP *)
BEGIN
  WRITELN('Plot Checking program; Pascal');
  WRITE('Number of data,NDATA?'); READLN(NDATA);
  WRITELN('Input pairs of X,Y values:');
  FOR I:=1 TO NDATA DO READLN(XLIST[I],YLIST[I]);
  MINMAX; (* Range of X and Y values *)
  PLOTP; (* Invoke plotter *)
END.
```

Printer plots from Fortran

The sample printer plotting subroutines coded in Fortran follow the pseudo-coed outline derived in L4.1. The Fortran subroutine follows very closely the structure of the Pascal procedure in the previous subsection. Indeed, I mostly transcribed the Pascal statements directly into Fortran statements. This had two advantages: First, the structure of Pascal forces logical development of programming steps, and this carries through into the Fortran coding. Second, after debugging the Pascal program (which, as an old Fortran buff, took me some time), it was straightforward to write a structured Fortran program that executes correctly.

One major difference between the Pascal and Fortran versions of the plotting program should be noted. This difference is in how some of the variables

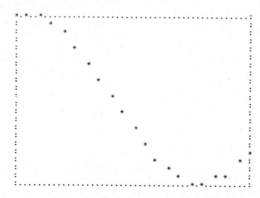

Figure L4.2.1   Sample execution of PLTCHK using 21 data points of the cosine function in the range 0 to 4.0 radians.

are shared between program segments. Pascal variables are, by default, shared between procedures, so that they tend to be "global" rather than "local." In Fortran, variables are local unless forced to be shared between routines by inclusion in argument lists or in COMMON blocks. I decided that use of COMMON is probably beyond the experience of most first-time users of the plotting program. Graphics programs using structured Fortran and COMMON are given in Appendix M of the textbook by Dyck, Lawson, and Smith.

The order of the program segments in the Fortran program is as follows: PLTCHK is the main program, which inputs the data and invokes the other subroutines. MINMAX determines the range of X and Y values. PLOTP generates the display and outputs it, following the pseudocode outline from L4.1.

The Fortran printer plotting program has been tested on the Apple II +, using Apple Fortran, and it has been tested on a mainframe computer using WATFIV, a structured Fortran described in the book by Dyck, Lawson, and Smith. The WATFIV version required input and output unit numbers, rather than the * for terminal input and output as shown here. The output from the Fortran version is nearly identical with that from the Pascal version, so it is not included.

```
      PROGRAM PLTCHK
C  Printer plots from Fortran; checking program
      INTEGER NDATA,I
      REAL XLIST(100),YLIST(100),XMIN,XMAX,
     1      YMIN,YMAX,XYFACT
      DATA NXGRID/50/,NYGRID/22/,MXDATA/100/
C  A * denotes terminal output or input
      WRITE(*,12)
   12 FORMAT(' Plot Checking program; Fortran')
      WRITE(*,14)
   14 FORMAT(' Number of data,NDATA?')
      READ(*,16) NDATA
   16 FORMAT(I2)
      WRITE(*,18)
   18 FORMAT(' Input pairs of X,Y values:')
      READ(*,20) (XLIST(I),YLIST(I),I=1,NDATA)
   20 FORMAT(2F5.2)
C  Find range of X and Y values
      CALL MINMAX(NDATA,XLIST,YLIST,
     1            XMIN,XMAX,YMIN,YMAX)
C  Invoke plotter
      CALL PLOTP(NXGRID,NYGRID,NDATA,
     1           XLIST,YLIST,
     2           XMIN,XMAX,YMIN,YMAX)
      END
```

```
      SUBROUTINE MINMAX(NDATA,XLIST,YLIST,
     1                    XMIN,XMAX,YMIN,YMAX)
C  Find limits in first NDATA elements of XLIST
      INTEGER NDATA
      REAL XLIST(100),YLIST(100),
     1     XMIN,XMAX,YMIN,YMAX
      XMIN = XLIST(1)
      XMAX = XMIN
      YMIN = YLIST(1)
      YMAX = YMIN
      DO 100 IXY = 2,NDATA
        IF (XMIN.GT.XLIST(IXY)) XMIN = XLIST(IXY)
        IF (XMAX.LT.XLIST(IXY)) XMAX = XLIST(IXY)
        IF (YMIN.GT.YLIST(IXY)) YMIN = YLIST(IXY)
        IF (YMAX.LT.YLIST(IXY)) YMAX = YLIST(IXY)
  100 CONTINUE
      RETURN
      END
      SUBROUTINE PLOTP(NXGRID,NYGRID,NDATA,
     1                   XLIST,YLIST,
     2                   XMIN,XMAX,YMIN,YMAX)
C  Printer plot of YLIST against XLIST
      INTEGER NXGRID,NYGRID,NDATA,IX,IY
      REAL XLIST(100),YLIST(100),XMIN,XMAX,YMIN,
     1     YMAX,XSCALE,XORG,YSCALE,YORG,XYFACT
C  Character default size is 1
      CHARACTER GRAPH(50,22),FIGURE,GROUND,
     1          XEDGE,YEDGE
      DATA FIGURE/'*'/,GROUND/' '/,
     1     XEDGE/'.'/,YEDGE/':'/
C  XYFACT is ratio of display scales
      XYFACT = FLOAT(NYGRID)/FLOAT(NXGRID)
C  X scale factor, then X origin
      XSCALE = (NXGRID-1)/(XMAX-XMIN)
      XORG = 1-XSCALE*XMIN
C  Y scale factor, Y origin
      YSCALE = (1-NXGRID)/(YMAX-YMIN)
      YORG = 1-YSCALE*YMAX
C  Make edges on GRAPH
      DO 200 IX = 1,NXGRID
        GRAPH(IX,1) = XEDGE
        GRAPH(IX,NYGRID) = XEDGE
  200 CONTINUE
      DO 300 IY = 1,NYGRID
        GRAPH(1,IY) = YEDGE
        GRAPH(NXGRID,IY) = YEDGE
  300 CONTINUE
```

```
C   Initialize rest of GRAPH to GROUND
        DO 500 IX = 2,(NXGRID-1)
          DO 400 IY = 2,(NYGRID-1)
            GRAPH(IX,IY) = GROUND
  400     CONTINUE
  500 CONTINUE
C   Locate display points, FIGURE, in GRAPH
        DO 600 I = 1,NDATA
C   X position
          IX = XSCALE*XLIST(I)+XORG+0.5
C   Invert Y position
          IY = XYFACT*(YSCALE*YLIST(I)+YORG+0.5)
          IF (IY.EQ.0) IY = 1
          GRAPH(IX,IY) = FIGURE
  600   CONTINUE
C   Output; loop over Y and X indexes
        DO 800 IY = 1,NYGRID
          WRITE(*,702) (GRAPH(IX,IY),IX=1,NXGRID)
  702   FORMAT(1X,50A1)
  800 CONTINUE
        RETURN
        END
```

The next subsection gives you many opportunities to use and modify the plotting programs just presented.

Plotting exercises

Three main types of exercises in using the plotting programs, either from Pascal or from Fortran programs, are suggested in the following: The first set of exercises requires you to prepare a variety of inputs, XLIST and YLIST, to the plotting procedures, but you do not have to modify the procedures themselves. The second group of exercises suggests small modifications to the plotting procedure to improve the picture, for example, modifying the figure and ground characters to the user's choice, or adding titles, scales, units, and tick marks. The third exercise set requires significant changes in the plotting procedure, especially for histograms or multiple figures with different symbols for each.

▶ *Exercise* L4.2.1   (Graphic examples)
(*a*) Generate as input to PLOTP a straight line. This can be used to indicate the resolution of the plots. It is also useful as a trial for such straight-line plots as you expect for a linear-least-squares fit, as used in computing laboratories L8 and L10.
(*b*) Use ellipses as input to PLOTP. The first ellipse that you use should be a circle, because you can use it to check the ratio of character x and y dimensions on your output device. After this, try an ellipse whose semi-major axis a and semi-minor axis b

in $x^2/a^2 + y^2/b^2 = 1$ are different. Beware! If you are clever and set XMIN $= -a$, XMAX $= a$, YMIN $= -b$, YMAX $= b$, with $a \neq b$, what will you get for your plot? Note that ellipses are to be plotted in the laboratory on space-vehicle orbits and trajectories, L11.

(c) Modify the input options to the main program so that the data to be plotted may be converted to semilogarithmic (on either axis) or to logarithmic on both axes as a user option. This conversion should be done before PLOTP is invoked. Test this with, for example, a function of the form $y(x) = \exp(ax)$, as encountered in the examples from the logarithmic century in A6.4.

(d) This exercise uses the computer to draw the complex plane discussed in A2. Suppose that the input consists of (x,y) pairs which represent the complex number $z = x + iy$. Assume that $|x| \leqslant 1$ and that $|y| \leqslant 1$. This determines the plotting scale. The program that invokes PLOTP should also generate the x and y axes for the complex plane plot. The axes can use the same figure character as the plotted points. □

The next set of exercises improves the pictures through small changes in the plotting program to allow more control by the program user and to include labeling of the plots. More familiarity with programming, especially for characters, is required in Exercise L4.2.2 than in Exercise L4.2.1.

▶ *Exercise L4.2.2*   (Improving your image)
(a) Modify PLOTP and its calling routine so that the difference between x and y display character dimensions may be taken into account by the program user. For example, if XYFACT is input different from zero, then use this value. If XYFACT is input as zero then choose some default value appropriate for the output device that you most often use.

(b) The ground character, GROUND, is currently set to a blank, and the figure character, FIGURE, is set to an asterisk, *. Modify PLOTP and the calling program so that the user may choose the characters to be used for ground and figure. For example, negatives (in the photographic sense) or "reverse video," as it is termed in the graphics game, may be very interesting. Note that if you use a very asymmetric character for FIGURE, such as T or L, this will worsen the resolution. Symbols that are of good character and resolution are *, =, -, I, O, H, +, X, 0, 8. Why are the lower-case equivalents of these alphabets not as good?

(c) Add information to the graph by allowing input of a title to be printed above the graph, by allowing axes labeled with the names of variables and their units, by the inclusion of scales on each axis, and by tick marks at the position of the scale values. (This option requires facility with output editing and formatting.) □

If you have done all of these exercises you will have a very elegant plotting package. You might now try extending your skills so that the type of figure to be displayed is variable. Part (a) can be done directly, but parts (b) and (c) require somewhat more modification of the basic program. They do not require that the previous two exercises have been done.

▶ *Exercise L4.2.3*   (A potpourri of plots)

(*a*) Modify the input to the main program PLTCHK so that it can accept pairs of values in polar coordinates, $[r,\theta]$, then convert them to Cartesian coordinates before invoking PLOTP.

(*b*) Modify the program PLTCHK and its procedures so that histograms can be generated. The user input should specify the width of each bin in the histogram, or alternatively, the number of bins in the histogram.

(*c*) Multiple figures on the same graph are often useful for comparison. Modify each program segment so that a series of XLIST, YLIST arrays can all be plotted on the same graph, each using a different figure character. The modifications consist mainly of incorporating the graph generating sections of the code inside loops that repeat for as many figures as you want to plot together. □

These exercises on printer plots show that a wide variety of simple graphs can be produced by this method. In the next section, other graphics techniques are briefly discussed.

L4.3 {2}   Other graphics techniques

In the preceding two sections we detailed the use of low-resolution graphics for output to a printer or to a video screen functioning in text mode. This section briefly surveys other graphics techniques, which require more-specialized output devices than does printer graphics.

The resolution of printer graphics using characters is limited by the character size and by the fact that there is usually unfilled space around each character. If instead of being generated by a solid type-element the characters are formed by selective firing of pins in a matrix on the printhead, the so-called "dot-matrix" method, then resolution of nearly an order of magnitude better can be obtained by selecting which dot to print, rather than which character to print. This dot-matrix method is very practical if you have such a printer. However, it is quite dependent on the computer hardware and software.

Video-screen and interactive graphics

The video screen that is used as a computer terminal is an ideal device for graphics, especially for interactive use and for animation. Typically the resolution for focusing the electron beam on the cathode-ray-tube screen allows about 400 points in the horizontal direction and about 200 points in the vertical direction. Indeed, the limitation is sometimes the computer memory capacity; for example, a 400×200 display with only an on-off capability and no intensity or color control for each display point requires 80,000 bits, which is typically 10 Kbyte in a small computer. If 7 intensity levels and 15 colors were available, then at least six times this amount of storage is needed just to hold the picture-generating information. Clearly, one picture is worth much more than a thousand words.

The basic concepts of computer graphics are covered in the primer by Waite. He discusses the hardware for a variety of graphics terminals, computers, and monitors for microcomputers. Waite's book has a detailed section on graphics for the Apple II computer using BASIC. The introductory book by Inman gives details of graphics for the TRS-80 microcomputer; the material on animated graphics is extensive and provides a good entry to this topic. The extension of computers to household use, and especially the use of computer graphics in these developments, is well described in the book by Fedida and Malik on the Viewdata/Prestel system, which is in use in many homes in Britain. Similar videotex systems are described in the book by Chorafas, and are in use in France, the United States, and the Netherlands.

The topics of projection, perspective, and stereoscopy in computer graphics are covered in a practical manner in the book by Angell. For the more advanced reader interested in computer languages for graphics, there are many texts and monographs. That by Giloi, which especially emphasizes data structures for graphics, algorithms, and languages, is especially suitable.

Static graphics

What do you do with a video graphics image that you want to keep? One possibility is to copy an image of the screen by photography. However, this tends to be cumbersome, requiring much extra equipment and usually producing only small pictures. Alternatively, a program can be used to reformat the display so that the contents of the graphics array can be output to a dot-matrix printer. This method is usually simple and inexpensive.

Some graphics programs and devices are intended primarily for static, hard-copy output. The programs are often included in program libraries for mini- and maxi-computers. Some canned graphics routines, such as Bell Laboratories' GR-Z for the Tektronix line of display terminals and plotters, can be used either on a video display or on a hard-copy output with very simple changes of the program parameters that invoke the graphics.

You will find an interesting survey of a wide range of graphics techniques and devices in the McGraw-Hill encyclopaedia article "computer graphics."

References on computer graphics

Angell, I. O., *A Practical Introduction to Computer Graphics*, Macmillan, London, 1981.

Bell Laboratories, *Documents for Use with GR-Z*, Murray Hill, N.J., 1977.

Bennett, W. R., *Scientific and Engineering Problem-Solving with the Computer*, Prentice-Hall, Englewood Cliffs, N.J., 1976.

Chorafas, D. N., *Interactive Videotex: The Domesticated Computer*, Petrocelli Books, New York, 1981.

Dyck, V. A., J. D. Lawson, and J. A. Smith, *Introduction to Computing*, Reston Publishing, Reston, Vir., 1979.

Fedida, S., and R. Malik, *The Viewdata Revolution*, Associated Business Press, London, 1979.

Giloi, W. K., *Interactive Computer Graphics*, Prentice-Hall, Englewood Cliffs, N.J., 1978.

Inman, D., *Introduction to TRS*-80 *Graphics*, Dilithium Press, Portland, Ore., 1979.

McGraw-Hill Encyclopedia of Science and Technology, "Computer Graphics," McGraw-Hill, New York, 1982.

Waite, M., *Computer Graphics Primer*, Howard W. Sams, Indianapolis, 1979.

L5   ELECTROSTATIC POTENTIALS BY INTEGRATION

This computing laboratory on numerical integration applies the formulas derived in section A4.4 to determining the pattern of electrostatic potentials arising from a charge distribution. The particular distribution suggested is that of a uniform density of charge along a straight line. This is chosen because the potential can also be obtained by analytic integration, as shown in L5.1.

Why bother with approximate integration methods, such as the trapezoid and Simpson formulas in L5.2 and L5.3, when an analytic formula is available? First, because the analytic result serves as a check case on the accuracy of these approximations. Second, because the numerical integration formulas can be used for charge distributions that are not analytically integrable. Indeed, if you write the integration routines in modular form, you will be able to use them in a wide range of applications. By having first compared them with algebraic results, you will have a good idea of their reliability.

Simple computer graphics, as introduced in L4, are suggested for use in L5.4 to display equipotential distributions. A variety of scaling options for such displays are also suggested in this section.

L5.1 {1}   Analytic derivation of line-charge potential

We consider the potential arising from a distribution of charge along a straight line. The most convenient coordinate system is as shown in Figure L5.1.1. The coordinate system is chosen to produce maximum simplicity and symmetry in the results. The distribution of charge along the line is written as $\lambda(x')$. In most of L5 we will consider $\lambda(x')$ to be independent of x'. The field point P at which the electrostatic potential is observed has coordinates (x,y).

The analytic derivation of the potential at P, $V(x,y)$, is made as follows. A small element of the line, of length $\Delta x'$, contributes an amount $\Delta V = \lambda(x')\Delta x'/r$, where r is the distance from the charge element to the field point, as shown in Figure L5.1.1. Clearly, this distance depends on x', which is what makes the problem interesting to compute. In the analytic derivation, we take the limit as $\Delta x' \to 0$ to produce the integral formula for the potential V observed at (x,y)

$$V(x,y) = \int dx' \, \lambda(x')/\sqrt{[(x - x')^2 + y^2]}, \qquad (L5.1)$$

in which the limits of integration are from $-a$ to a, the region over which the charge is nonzero. In this equation we have also given the explicit expression for r, obtained using the theorem of Pythagoras.

In the numerical approximation to the line-charge potential we will keep $\Delta x'$

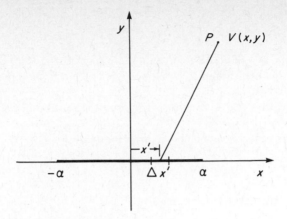

Figure L5.1.1   Coordinate system for a line-charge distribution extending from $-a$ to $+a$.

finite but sufficiently small so that the error involved in the numerical integration approximations will be satisfactorily small.

► *Exercise* L5.1.1   The smallest unit of free electric charge is probably the charge
on the electron or proton. The magnitude of this charge is $e = 1.6\times10^{-19}$ Coulomb.
Suppose that the charge density $\lambda = 10^{-6}$ Coulomb/meter. What is the average
spacing of the elementary charges? Under what conditions is it then sufficiently
accurate to consider the charge distribution as continuous, and to take the limit
$\Delta x' \to 0$? □

Electrostatic potential of a line charge

If the charge density $\lambda(x')$ in (L5.1) is constant, then it can be factored out of
the integral, and simply called λ. Further, if $y \neq 0$, we can change variables to

$$z = (x - x')/y, \tag{L5.2}$$

to get a simpler integral

$$V = \lambda \int dz/\sqrt{(z^2 + 1)}, \tag{L5.3}$$

in which the limits of integration are $(\pm a - x')/y$. The indefinite integral is a
standard form $-\ln |\sqrt{(z^2 + 1)} - z|$, as can be verified by differentiating this
result. Thus, the potential at the point (x,y) due to a uniform line charge along
the x axis from $-a$ to a is

$$V(x,y) = \lambda\ln\{[\sqrt{[(x - a)^2 + y^2]} - (x - a)]/[\sqrt{[(x + a)^2 + y^2]} - (x + a)]\}. \tag{L5.4}$$

This result holds provided that the field point (x,y) is not along the charge distribution, that is, either $y \neq 0$ or $|x| > a$.

▶ *Exercise* L5.1.2   (Checking the analytic integration)
(*a*) Show the details leading to equation (L5.4) for the potential, including evaluation of the integral.
(*b*) Verify equation (L5.4) by direct differentiation of the indefinite integral for the potential.
(*c*) What aspects of the derivation restrict its validity? ☐

On-axis potential of a line charge

We now consider the special case in which all the charge is viewed end-on, that is $y = 0$. The derivation of the potential must now be modified, because the substitution (L5.2) is no longer valid. The potential integral reduces to

$$V(x,0) = \lambda \int dx'/|x - x'|, \tag{L5.5}$$

in which the range of integration is from $x' = -a$ to $x' = a$. If $|x| > a$, then the field point $P = (x,0)$ is not within the charge distribution. Otherwise, since $y = 0$, when $x' = x$ the potential will become indefinitely large (within our approximation that the charge distribution is of zero thickness.)

Ignoring this shocking possibility, the integral in (L5.5) can be evaluated in a straightforward way. It gives the analytic result for the on-axis potential of a uniform line charge

$$V(x,0) = -\lambda \ln [(|x| - a)/(|x| + a)]. \tag{L5.6}$$

When you evaluate the integral, the two cases $x > a$ and $x < -a$ should be considered separately. However, the result (L5.6) includes both cases.

▶ *Exercise* L5.1.3   Show in detail the steps involved in the integrations of (L5.5) for the on-axis potential. ☐

For $y = 0$ the invariance of the potential $V(x,0)$ under the symmetry operation of reflecting the x coordinate through the origin, $x \to -x$, is evident from the formula. Does this symmetry also hold for $y \neq 0$?

Symmetries and reflections

The purpose of the following remarks is to illustrate how the use of symmetry principles can simplify the calculation and understanding of physical systems. Eugene P. Wigner was one of the pioneers of modern science in the development of symmetry principles, especially through the use of the mathematical theory of groups. A collection of his essays, *Symmetries and Reflections* (Indiana University Press, Bloomington, 1967), makes very interesting reading.

The application of reflection symmetries to the potential from a uniform line charge is made as follows. From Figure L5.1.1 we see that our choice of coordinate system and the reflection symmetry of the charge distribution should lead to the result that the potential is the same at corresponding points in each of the four quadrants. Is this expectation fulfilled by the result (L5.6)? If we rationalize the denominator of this expression by multiplying numerator and denominator of the argument of the logarithm by the numerator, we find after a little algebra, that for $y \neq 0$ the line-charge potential can also be written as

$$V(x,y) = \lambda \ln\{[\sqrt{[(x-a)^2 + y^2]} - (x-a)][\sqrt{[(x+a)^2 + y^2]} + (x+a)]\}. \quad (L5.7)$$

▶ *Exercise* L5.1.4   Do a little algebra as a change from computing, by verifying this symmetrized form of the line-charge potential. □

We have now reduced the potential formula to one that is manifestly invariant under $x \rightarrow -x$, and to $y \rightarrow -y$. That is, the potential is the same in all four quadrants and depends only on the absolute values of x and y.

Why bother with this use of symmetry? First, it provides a check on the correctness of the formula from our geometric and physical intuition. Second, the potentials need be calculated and displayed in only a single quadrant, which provides an immediate saving of a factor of four in computing effort, and a factor of two increase in linear resolution when the potential is displayed.

Scaling the line-charge potential formulas

The line-charge potential formulas can be further simplified if we invoke scaling symmetries. This also avoids potential problems in the choice of electrical units. The scaling is done as follows. If all lengths in the potential problem are expressed as multiplies of a, as

$$X = x/a, \quad Y = y/a, \quad (L5.8)$$

and if the potential is expressed as a multiple of λ, as

$$V_s = V/\lambda, \quad (L5.9)$$

then we get scaled versions of the potential formulas (L5.6) and (L5.7). For $Y \neq 0$

$$V_s(X,Y) = \ln\{[\sqrt{[(X-1)^2 + Y^2]} - (X-1)]/[\sqrt{[(X+1)^2 + Y^2]} - (X+1)]\}, \quad (L5.10)$$

and for $Y = 0$

$$V_s(X,0) = -\ln \{(|X| - 1)/(|X| + 1)\}. \tag{L5.11}$$

▶ *Exercise* L5.1.5   Derive these scaled versions of the line-charge potential formulas. ☐

We notice that scaling of the lengths and of the potential has reduced the parameters in the problem from x, y, a, λ to X, Y, which are the essential variables of the problem. Note also that X, Y, and V_s are dimensionless quantities. Therefore we do not have to label any output with appropriate units, thus avoiding the imprecations of punctilious pedagogues. However, practical application of these formulas certainly requires actual units.

Scaling of dimensions, such as x and y, and of observables, such as the charge density λ and the potential V, to appropriate units, also reduces the chance that very large (or very small) numbers will appear in the calculations. Since each computer system has only a finite range of the magnitude of the numbers that can be represented in it, such *overflows* (from too large numbers) or *underflows* (from too small numbers) can be minimized by appropriate scaling. Sometimes this type of error is not even diagnosed by the computer, leading often to meaningless results.

L5.2 {1}   Line-charge potential using trapezoid formula

The programming exercises in this and the next section require the writing of several procedures, each of which calculates the potential due to a uniformly charged line. These procedures are: (1) Analytic evaluation of the scaled potential using formulas (L5.10) and (L5.11). (2) Numerical integration of the potential using the trapezoid formula from A4.4. (3) Simpson's formula from A4.4 for the numerical integration of the potential. (4) Generation and graphing of equipotential curves. We outline the structure of the program in the following.

Structure of the electrostatic potential program

The following is a pseudocode program outline that you may use as an outline for your calculations of electrostatic potentials.

L5 Electrostatic potentials by integration

Data and options input

L5.1 Analytic formulas for the potential:
If $Y \neq 0$ calculate $V_s(X,Y)$
If $Y = 0$ and $|x| > 1$ calculate $V_s(X,Y)$

Output analytic potential
Option to generate array of $V_s(X,Y)$
L5.2 Line-charge potential using trapezoid formula:
    Trapezoid integration formula
        Initialize integral to average of end-point values
        Sum in loop over intermediate points
        Multiply sum by step size, then output
L5.3 Potential integration using Simpson's formula:
    Simpson integration formula
        Initialize integral to sum of end-point values
        Sum even intermediate points
        Sum odd intermediate points
        Multiply sums by coefficients and step size, then output
L5.4 Displaying equipotential distributions:
    Find minimum and maximum potentials
    Generate and output linear display
    Convert linear display to other options

Go for more data and options

Program end.

For convenience of organization, we divide the task of writing these procedures into a series of exercises.

▶ *Exercise L5.2.1* (Analytic evaluation of the potential)
(*a*) Write a program segment that inputs the scaled coordinates X, Y of the field point at which the potential is observed, calculates the scaled line-charge potential, V_s, at that point, then outputs X, Y, and V_s. The program segment should then go for more input.
    A meaningful way to terminate execution of the analytic formula segment is to input values of the scaled coordinates for which the potential is undefined, namely $|X| \leqslant 1$, $Y = 0$. A more instructive way would be to output an error message to the program user if such coordinates are input, but to terminate only if the observer is put in the middle of the line charge, at 0, 0. If you are programming in Fortran, it would be practical to use the statement function to define the potential formula.
(*b*) Check the symmetry of the line-charge potential formula by evaluating it for representative values of X, Y then $-X$, Y, X, $-Y$, and $-X$, $-Y$. If you are convinced of these reflection symmetries, then you might confine future calculations to the first quadrant.
(*c*) On a sheet of coarse graph paper record the scaled potential at a few points in the first quadrant, such as close to the charge distribution and as far as 4 units away from it. These values are presumably exact, within the error of evaluating the natural logarithm function in (L5.10) or (L5.11) and the round-off or truncation error. Therefore, these values will be useful as a check on the accuracy of the numerical

integration algorithms. To record the values on a coordinate grid, rather than as a list, provides more insight into the results.

(*d*) At large distances from the line charge the potential should decrease as $1/r$, where $r = \sqrt{(x^2 + y^2)}$ is just the distance from the center of the charge. Check this prediction by using large X and Y values, especially combinations that keep the distance fixed. The equipotentials should lie approximately on circles centered on the midpoint of the line charge. You might add some code to calculate the distance and to output it. □

The analytic formula program in this exercise will have generated test cases with which to compare the numerical integration results. We now suggest a program outline for the trapezoidal rule.

▶ *Exercise* L5.2.2   (Trapezoid program for line-charge potential)

(*a*) Write and debug a program segment that evaluates an integral by using the trapezoid formula (A4.26), with a loop structure to step through the values of x. There are NTRAP such values. A useful test function is the linear form $y(x) = x$. If you have correctly coded the trapezoid formula, including the special treatment of the endpoint values $y(a)$ and $y(b)$, then it should give exact results, apart from round-off errors. Since the integral of x from -1 to 1 is zero, this provides a good test case.

(*b*) Modify the input to your potential program so that in addition to the field-point coordinates X, Y the number of points to be used in the trapezoid formula is also input. Call this NTRAP. Note that the number of intervals into which NTRAP points (including endpoints) divides a line of length 2 (in our scaled units) is NTRAP $-$ 1. (This result is also important in counting the number of pickets for a fence.) Thus, the step-size to be used in (A4.26) is $h = 2/(NTRAP - 1)$.

The function to be integrated is just the scaled potential contribution from an element of charge at a point on the line of charge. Therefore, in the trapezoid formula (A4.26) we have, on comparison with (L5.3) appropriately scaled,

$$y(x) = 1/\sqrt{[(X - x)^2 + Y^2]}, \qquad (L5.12)$$

and the range of integration is from -1 to 1. For a Fortran program it's a good idea to make this a statement function, in order to separate the trapezoid formula from the details of specifying the function being integrated.

(*c*) Vary NTRAP for representative values of X, Y. Explore how the potential estimated by using the trapezoid formula for the integral varies as NTRAP increases. Modify your program so that a comparison between the analytic and trapezoid formula results is made. A good indication of the discrepancy is an output of the fractional difference (including sign) of the two results. Your output should be clearly labeled so that the sign of the difference is meaningful. Is the dependence of the trapezoid formula error on step-size h consistent with that indicated in A4.4?

(*d*) This part of the exercise makes a beginning in the display of equipotentials. To do this, modify the input instructions to read the range of X and Y values, say XMIN to XMAX by DX, and YMIN to YMAX by DY. Then fill an array with values of the scaled potential by putting the trapezoid formula (or analytic formula) inside a double-nested loop over the values of X and Y. At the bottom of this nested loop, code output statements that produce a table of the scaled potential. The table should be labeled on the left by the Y values, and on the bottom by the X values. The number of significant figures should be no more than the precision of the calculation, as determined in (*c*).

On your printout you now have a rough way of drawing the equipotential curves by merely connecting the coordinates at which the potentials are nearly equal. Section L5.4 suggests a more elegant method of displaying equipotentials. Note that to produce the Y axis with the conventional orientation the Y index of the array has to be printed out backward. (Or, you could code the Y loop so that it was decreasing rather than increasing. This is straightforward in Pascal, but less direct in Fortran.)
□

This computing exercise completes the investigation of the trapezoid formula for numerical integration. You may make further use of the trapezoid algorithm in L10 for calculating Fourier transforms.

L5.3 {1}   Potential integral from Simpson's formula

The discussion of Simpson's formula in A4.4 showed that it could be much more accurate than the trapezoid formula because it approximates segments of curves by parabolic arcs, rather than by straight-line segments. Thus, the curvature of the integrand is taken into account, and the error in the integration depends on the step-size h as h^5, rather than as h^3 in the trapezoid formula.

To compare the two formulas, the exercises programmed for the trapezoid rule can be repeated for Simpson's rule.

Simpson's formula exercises

The following set of programming exercises require small modifications to the structure of the trapezoid formula program in Exercise L5.2.2. The purpose of this exercise is for you to appreciate the improvements in numerical accuracy and efficiency that can be achieved by relatively simple changes in algorithms and programming.

► *Exercise* L5.3.1   (Simpson program for line-charge potential)
(*a*) Write and debug a program segment for the Simpson formula (A4.29). The parabolic function $y(x) = x^2$ integrated from -1 to 1 should be integrated exactly by the Simpson formula, giving a value of 2/3.
(*b*) How does the accuracy of the Simpson formula depend on NSIMP, that is, on the step-size h? Is the dependence consistent with the error estimate in (A4.4)?
(*c*) Modify your program input so that the number of points for the Simpson formula, NSIMP, is also input. Note that NSIMP must be odd (even number of inter-

vals), which you can achieve by having your program compute, output, and use the nearest larger odd integer if necessary. The function to be integrated by the Simpson formula is given by (L5.12). Compare the accuracy of these results with those from the trapezoidal rule (Exercise L5.2.2) and the analytic integral (Exercise L5.2.1). Do this for an interesting range of X and Y values. □

These exercises on numerical integration have been illustrated with the trapezoid and Simpson rules and applied to determination of the potential arising from a finite line of uniformly distributed electric charge. It is now of interest to consider meaningful ways of presenting the results of the potential calculations.

L5.4 {2}   Displaying equipotential distributions

The simple tabulations of potential values that you have made in the preceding exercises will, I hope, have convinced you that numerical output is not very suitable for giving an overall picture (!) of a potential distribution.

Why are numerical values of the potentials often unsuitable? Probably because humans perceive mainly by pattern recognition, for which the task of recognizing and comparing strings of numerical symbols is very difficult. (''Computers quantify, humans recognize.'') The output from the calculation of equipotentials is therefore more meaningful if the numerical values are converted into much simpler display patterns. Familiar examples are weather maps, showing lines of equal pressure (isobars), streamlines in nonviscous fluids, lines of equal temperatures (isotherms), not ignoring isotacs and isotachs. The following discussion and sequence of exercises make suggestions for such a display.

Equipotentials of a line charge

As the field of values to be displayed in contour form we use equipotentials from calculations of the potential of the uniform line charge considered in sections L5.1 through L5.3. This is convenient rather than necessary. With care in program design you can make the generation of the display array a separate program segment from that for the equipotential calculations and their output. Thus your contour line program will be of general utility.

▶ *Exercise* L5.4.1   (Programming equipotentials)
Generate an array of the scaled potential $V_s(X, Y)$ for scaled coordinate values $X > 0$, $Y > 0$ only, with the restriction $X > 1$ for $Y = 0$. (You might insert special values at the points corresponding to the position of the line charge, $X \leqslant 1$, $Y = 0$, so that the display routine marks the position of the line charge.)

It is probably most efficient to use the analytic formula for the potential, rather than the numerical integration approximations, especially since the display on a printer-type device has rather coarse resolution, which would make the difference between the two not resolvable. (However, an interesting display would be the *differences* between the analytic and numerical results; the scaling procedure indicated in the following will take care of the fact that the differences will be relatively small.)

A fairly large array of (X, Y) points, say (40, 40), is desirable for the equipotential display, in order that fairly smooth contours be evident. The symmetries of this particular potential, derived in L5.1, require only the first quadrant $X > 0$, $Y > 0$ to be shown. With these preliminaries, there are various options for the display:

(*a*) Write and test a procedure to find among all of the $V_s(X, Y)$ values the smallest, VMIN, and the largest, VMAX. This is best done by a nested loop structure in which the loops are over X and Y. A sample procedure, MINMAX, is given in L4.2. The values VMIN and VMAX are needed for scaling in the remainder of this exercise.

(*b*) The first display is linear and outputs integer values in the range 0 to 9 for the potential range VMIN to VMAX. To do this, consider an integer array IV whose (IX,IY) element corresponds to the potential $V_s(X, Y)$. For the linear display the correspondence is established as follows:

$$IV(IX,IY) = 9(V_s - VMIN)/(VMAX - VMIN). \qquad (L5.13)$$

The array IV should then be output with IX indicating the columns and IY indicating the rows. (For a printer display the smallest values of Y should be output first, as discussed in L4.1.)

(*c*) The next display method makes a more distinct pattern between successive equipotential values in the linear display. Using the integer elements from the array IV calculated in (*b*), assign a + sign to the even elements and a blank to the odd elements. This requires an array of character type; call it PB, for Plus or Blank. Within a double-nested loop over IX and IY we could have the statement (in pseudocode)
  If IV(X,Y) = 2*(IV(X,Y)/2) then PB(IX,IY): = ' + ' else PB(IX,IY): = ' '
(In Fortran, depending on your local dialect, this may require some use of DATA statements for allocating the values of the characters + and "blank.") This display will have the advantage that successive contours are readily distinguished, but the disadvantage that relative values are lost.

A more elegant compromise would be to display the even values but to blank out the odd values. This would require transforming the integers 0 through 9 to the characters '0' through '9'. Suggestion: Write a small procedure to convert integers to their character representation.

(*d*) Non-linear scaling of the display contours can be made in order to selectively enhance some parts of the display. For example, a logarithmic scale enhances low values over high values. (Detail on logarithmic displays is given in the following subsection.)

In computer applications a binary display scale, that has a convenient relation to base 2, the basis of binary arithmetic, is very suitable because it is directly related to the digital computer representation of numbers. For example, if a scale from $000_2 = 0$ to $111_2 = 7$ is used, then only 3 binary bits are required to store each display value. If a sign bit is also needed, then a total of 4 bits is used. This is very convenient for most machine-level tasks, although the resolution (8 levels) is fairly coarse.

The appropriate scaling for a 3-bit binary display is provided by allocating an integer array, say LG2, and making the transformation

$$LG2(IX,IY) = 7lg_2[(V_s - VMIN)/(VMAX - VMIN) + 1]. \qquad (L5.14)$$

This can then be displayed on an integer scale of 0 to 7, as in (*b*), or transformed to the + and blank representation, as in (*c*), before the display is made. □

Using logarithmic scales for displays

There are several justifications for the logarithmic displays suggested in part (*d*) of the previous exercise. Among the reasons are the following: (1) A logarithmic display selectively enhances relatively small values, which are often the interesting ones in research problems. (2) Computer storage and manipulation is simplest with binary scales. (3) Many phenomena change with time in an exponential manner, as emphasized in A6.4 on the Logarithmic Century. (4) Human sensory response, especially hearing and sight, are very roughly logarithmic with respect to intensity; otherwise we could not appreciate the rustle of spring and withstand the roar from a jet turbine. Logarithmic sensory response is also related to the use of decibel scales, as discussed in section A6.4.

A color-coded display is often most appropriately shown on a logarithmic scale. For example, topographic maps often used compressed scales for high elevations and oceanic depths and appropriately color code these from white, brown, green, to deep blue. The 8-level logarithmic display transformation given by (L5.14) is suitable for microprocessor systems with color graphics, such as the Apple II.

References on a variety of graphics techniques for computers from micros to mainframes are provided in the reference section of the introduction to computer graphics, computing laboratory L4.

L6   MONTE-CARLO SIMULATIONS

In this computing laboratory you will have several opportunities to investigate numerically the effects of randomness in phenomena of the natural sciences. Also, there are several examples from mathematics, including the drunkard's walk problem, round-off error in arithmetic, and geometry by random numbers.

The technique of Monte-Carlo simulation is so-called because of the element of chance, obtained by the use of random numbers, in the simulation calculations. Since it is important that no special bias be introduced in using the random numbers, we discuss at length in L6.1 the generation of random numbers. Section L6.2 offers several examples and exercises of Monte-Carlo simulations applied to problems in arithmetic, statistics, and geometry. Did you know that there are several ways of estimating π by Monte-Carlo methods?

Modeling the real world by Monte-Carlo simulations is the emphasis in L6.3 and L6.4. In section L6.3 you can investigate the approach to thermal equilibrium and the increase in entropy of a gas allowed to expand into an initially empty container. The simulation of nuclear radioactivity can be studied by simple Monte-Carlo methods in L6.4 and compared with the exponential decay laws established experimentally. This chapter concludes with a discussion of other simulation examples and references on Monte-Carlo methods.

L6.1 {1}   Generating and testing pseudo-random numbers

Computer simulation experiments, as they were discussed in L1, often require numbers taken at random from a given range. Such numbers may be used to simulate numerically processes that either are believed to be truly random or whose complicated inner workings we ignore. For example, in their study of the settlement of Polynesia by drift voyages, Levison, Ward, and Webb simulated random ocean current and wind directions by using random numbers. As you may read in their book, they found that New Zealand, one of the largest and most temperate island groups in the South Pacific, could probably not be settled by random drift voyages.

One method of generating numbers that are probably random is to take a natural phenomenon having an intrinsic element of randomness. For example, the time at which a given nucleus decays in a sample of radioactive material is believed to be randomly distributed about a mean value. Indeed, tables of random numbers have been generated by this procedure, as is described in the article by Frigerio and Clark.

For computer applications it is most convenient to use arithmetic operations in the computer to generate the random numbers as they are needed. One method is to take two large numbers describing (let us hope) independent computer operations (for example, the time of day in milliseconds and the current contents of some workspace element), combine them in a complicated way, then extract the required number of digits from the results. Such a method depends strongly on the computer used, and it is impractical when testing a program that uses random numbers, because the numbers are not reproducible.

An alternative computer method of obtaining fairly random numbers is to use randomizing arithmetic operations applied to a sequence of numbers. If the rule is complicated enough, and the numbers large enough, we may hope that some of their digits will occur in a random sequence. One calls numbers generated in this way "pseudo-random numbers." However, for brevity, I will usually call them just random numbers. A common generator, which works reliably if used with skill and caution, is the power-residue generator, which we now describe.

Power-residue random-number generator

The formula for generating random numbers by the power-residue method relates the nth number, r_n, to the $(n + 1)$th number, r_{n+1}, by

$$r_{n+1} = \text{mod}\,(cr_n, M). \tag{L6.1}$$

In this formula for generating random numbers c is the multiplicand, and M is the modulus with respect to which the remainder, which is r_{n+1}, is calculated. The value r_1 is called the *seed* random number. It must be supplied before the generator can begin. (Finding a seed is easy if you have an Apple.) Each quantity in (L6.1) is an integer, and integer arithmetic must be used.

As an example of the arithmetic in the power-residue method, consider $c = 3$, $r_1 = 5$, $M = 13$. Then we obtain $r_1 = 5$, $r_2 = 2$, $r_3 = 6$, $r_4 = 5 = r_1$, and so on, in a cycle of length 3. Note that there are M integers in the range 0 to $M - 1$, so that the period can not be longer than M. In fact, in this example we obtained a period of 3, which is much shorter than the maximum period of 13. A zero is a fatal integer to obtain for the remainder, because you can thereafter never get away from it if (L6.1) is used to generate the random numbers.

▶ *Exercise* L6.1.1   In this power-residue random number generator change only the multiplicand c, to $c = 11$. What is now the period of the random-number sequence? ☐

Clearly, large values of c and M are necessary in the pseudo-random generator in order that the digits in the remainder become scrambled, thereby increasing the chance that the sequence does not repeat so quickly.

For the random-number generator (L6.1) recommended values (as discussed in the book by Knuth), are: for the multiplicand $c = 8m + 3$, where m is a positive integer and c of order $\sqrt{M}$; for the seed r_1 an odd integer of order $\sqrt{M}$; for the modulus M a large odd integer.

It is often stated that the modulus M should be the largest integer that the computer can handle. However, this can get you into trouble with overflow during the calculation of r_{n+1} by formula (L6.1). To see this, note that r_n can be as large as $M - 1$, so that if $c \approx \sqrt{M}$ then the product cr_n greatly exceeds M, and thereby the range of integers. For this choice of c the value of M should be of order the *two-thirds power* of the largest integer that the computer can handle. For example, on a small computer the largest integer may be 32767 ($2^{32} - 1$), so that M should be about 10^6 ($\approx (2^{32})^{2/3}$). On a large scientific ("mainframe") computer the largest integer may be just less than 2^{64}, so that M as large as 2^{42} would be appropriate.

▶ *Exercise* L6.1.2   Why should even integers be avoided in generating pseudo-random numbers by the power-residue method? □

The generation and testing of random numbers is described in great detail in the handbook by Dudewicz and Ralley.

Random numbers in a given range

We often want random numbers in a range different from the 0 to $M - 1$ provided by (L6.1). If we want a sequence of random numbers in the range s_- to s_+ then we use

$$s_n = s_- + (s_+ - s_-)r_n/M, \qquad (L6.2)$$

with each integer r_n obtained as in (L6.1). As an example of the use of (L6.1), to get random numbers uniformly distributed in the range -1 to 1 we use $s_n = (2r_n/M) - 1$. Note that one must be careful in the mixed-mode arithmetic that the division in (L6.2) is done on the real representation of r_n, rather than on its integer representation.

▶ *Exercise* L6.1.3   (Program for array of pseudo-random numbers)
(*a*) Write a program that inputs the values of the REAL variables s_- and s_+, and of the INTEGER variables c, r_1 and M, and the number of random numbers required, NUMRAN. Your program should then check that the array size to be used to store the random numbers is not less than NUMRAN. The program should then enter the program segment that generates an array of random numbers. This segment could be a library program, or you could code it yourself in part (*b*) of the exercise.
(*b*) Write and debug a program segment that generates an array of NUMRAN random numbers in the range s_- to s_+ using equations (L6.1) and (L6.2).
(*c*) Make a frequency histogram display about 10 bins wide. This could be done by hand, or you could use the display program from L4. What did you expect for the frequency histogram? What did you get and are the differences between expectation and reality significant?

(*d*) Could it ever be proved that the sequence that you have produced is truly random? ☐

The random walk

In part (*c*) of the previous exercise you were asked to estimate the level of significance of deviations between expected and observed distributions of random numbers. From your experience with statistics you might have guessed that a sample of size N, for a reasonably large sample size, has a standard deviation of about $\sqrt{N}$. This is indeed a fair estimate. To illustrate this we consider the problem of the random, or drunkard's, walk. The problem is posed as follows: A drunkard stumbles from a lamp-post over a flat surface, taking steps of roughly equal size but in random directions for each step. After taking N steps, how far is the drunkard expected to be from the lamp-post?

We give a vigorous physical demonstration of the result, rather than a rigorous mathematical proof. After N steps the total distance, D, from the lamp-post is, by Pythagoras, obtained from

$$D^2 = (x_1 + x_2 + \dots + x_n)^2 + (y_1 + y_2 + \dots + y_n)^2, \qquad (L6.3)$$

where each x and y value is randomly distributed, but the RMS displacement is one step. On expanding the squares of the sums, there are sums of squares, all of which are positive, and sums of cross products between the x values and similar sums of cross products between different y values. But these cross products have random signs, so they will tend to sum to zero for N large enough. Thus, only the sum of squares remains, so that

$$D \approx \sqrt{N}, \qquad (L6.4)$$

where D is in units of steps. For example, after 900 steps the drunk will have gone only about 3 times as far from the lamp-post as after the first 100 steps.

Brownian motion of small objects in fluids is another example of a random walk, in which the randomly directed molecular forces gradually displace the object. (Liquids other than alcohol are quite suitable for displaying the effect.) A Monte-Carlo simulation of two-dimensional Brownian motion, suitable for programming on a microcomputer, is presented in the article by Mishima et al.

▶ *Exercise* L6.1.4   (Steps in the random walk)

(*a*) For the random-walk derivation show the algebraic steps from (L6.3) to (L6.4) and present the randomness arguments in detail.

(*b*) Use the random-number program segment from Exercise L6.1.3 to generate x and y values in the range $-1/\sqrt{2}$ to $1/\sqrt{2}$ then to compute the distance D using (L6.3). At each step, output D and $\sqrt{N}$. How quickly does $D \rightarrow \sqrt{N}$? ☐

It is important, but unfortunate, to note that in such random-walk and other Monte-Carlo problems the convergence is relatively slow, improving only as $1/\sqrt{N}$. This behavior will be familiar to you if you have counted pulses from nuclear radioactivity or from discrete photon events in optics. It takes 4 times the number of counts to improve the statistical reliability by a factor of 2. This can be investigated in the exercises in L6.4 on the Monte-Carlo simulation of nuclear radioactivity.

L6.2 {1}   Stimulating simulations in mathematics

In this section we consider several introductory examples and exercises in the use of random numbers and Monte-Carlo simulation applied to mathematics. These examples are; the simulation of round-off error in numerical computations and the estimation of areas using Monte-Carlo methods.

Simulation of round-off errors

In sections A4.2 and A4.3 and in Problem [A4.1] we discussed numerical noise, round-off, and other error sources in computations with a finite number of significant figures. A practical way of experimenting with the effects of numerical noise is to simulate the round-off error by using a random-number generator to include an error that may be much larger than that inherent in the computer being used. Furthermore, the size of the round-off error is under your control. For example, starting with random numbers in the range 0 to 1, you can produce an error in the dth significant figure, with both random magnitude and sign, by multiplying each number in the nth step of a calculation by $[1 + 10^{-d}(2r_n - 1)]$.

▶ *Exercise L6.2.1*   (Random numbers and arithmetic error)
(*a*) Write a simple program that simulates round-off error buildup in the computation of the square root. Here is an outline in pseudocode:

```
Round-off error in SQRT(X) after N steps with D significant figures
    Input X, N, D; save X in XSAVE
    Loop over n for N times
      X = SQRT(X)
      Get a random number RN in range 0 to 1
      then modify X in the Dth significant figure
      X = X (1 + 10⁻ᴰ (2*RN − 1))
      Square X to reverse the SQRT operation, within the loop
      X = X*X
    End of loop over N
    Output XSAVE, N, D, X, and ERR  = XSAVE − X
    Go for more data and options
  Program end.
```

The values of X that you choose should be surd (irrational). For large enough N, say N > 100, is ERR roughly proportional to $\sqrt{N}$, as our random-walk model claims? Cautionary note (or an invitation to explore): If you make D about the same as the number of significant figures carried by the computer, then you will produce strange effects from the interplay between the computer's round-off and your random-number simulation of round-off.

(*b*) Compare the effects of round-off with those of truncation by replacing the random-number multiplication step in the preceding program by $X = 10^{-D}\text{INT}(10^{D}X)$, where the function INT takes the integer of a decimal number. Repeat the calculations with similar values of X, N, and D as in (*a*). In truncation, does the error ERR increase roughly proportional to N, as Problem [4.1] suggests? □

This exercise completes our simulation of round-off in computer arithmetic. The next mathematics application of Monte-Carlo simulation is simple, but its extensions to more-complicated situations are of great importance in applied science.

Estimating integrals and areas

As a very simple example of the use of Monte-Carlo methods, we consider calculating areas and integrals. The example, which leads to an estimate of π, is so simple that understanding how the simulation works is as easy as pie.

Consider the arc of a circle located in the first quadrant, as shown in Figure L6.2.1. We know that

$$\int dx \int dy = \pi/4 \tag{L6.5}$$

if the double integral is over the arc of the unit circle located in the first quadrant. Suppose that we had forgotten how to carry out this integral, which is rather tricky in Cartesian coordinates. We could simulate the determination of the area by sampling points at random within the dashed square. The probability that a sampled point will lie within the arc tends to (for a large number of samples) the ratio of the area enclosed by the arc to that enclosed by the dashed line. This ratio is just $\pi/4$.

► *Exercise* L6.2.2   (Estimating π by Monte-Carlo simulation)
(*a*) Inscribe an arc of a circle in a square with sides of about 50 cm length. Drop from about 1 m height handfuls of pebbles from above the figure but from a fairly random position. Does the fraction of the pebbles that come to rest inside the circle to those that stop inside the square tend to $0.79 \approx \pi/4$ after you have dropped many pebbles? Note that this analog computer could have been used long before Euclidean geometry was developed, but who would have been interested in it?
(*b*) Write a small computer program that generates pairs of random numbers *x,y*, with each number in the range 0 to 1, so that the points lie in the first quadrant.

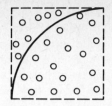

Figure L6.2.1   Circular arc (solid) and square boundary (dashed) for estimating π by Monte-Carlo simulation. The points indicate randomly chosen, but uniformly distributed, sampling points. The estimate shown is $4 \times 17/(17 + 7) = 2.8$.

Increment a counter, TOTAL, by 1 for each (x, y) pair. If $x^2 + y^2 < 1$, then increment a counter, CIRCLE, by 1. After many such trials does the ratio CIRCLE/TOTAL tend to $\pi/4$? Does the deviation decrease roughly as $\sqrt{N}$, where N is the number of trials?

(c) (Count Buffon's needles[1]) Needles are dropped at random on a plane surface ruled parallel with lines whose spacing is equal to the length of the needles. Prove that the probability that a needle intersects a line is $2/\pi$. See Gamow, pages 219–222.

(d) (Counting Buffon's needles). Do the experiment described in part (c) either by dropping needles, matches, chopsticks, or such, or by writing and running a Monte-Carlo simulation program in which the random variables are the distance from the ends of the needles and the angle between the needles and the lines. An experimental determination of π by similar methods to Buffon's needles simulation was reported in the article by Hall published in 1873. □

This completes our examples of Monte Carlo simulations in mathematics. Most electronic games, especially those based on geometry, have randomness included in their programming, thus ensuring that the players will probably keep on trying their luck. Engel's book contains several examples of such stimulating simulations.

L6.3 {2}   The approach to thermodynamic equilibrium

The next two sections provide extensive applications of Monte-Carlo simulation to modeling real-world systems. Both examples, the approach to thermodynamic equilibrium and nuclear radioactivity, can also be modeled as continuous processes approximated fairly adequately by differential calculus methods. This enables us to compare the results of the simulation method with the analytic results that we derive by calculus techniques.

[1]Among other interests of George Buffon (1707–1788) were natural history, and he anticipated Charles Darwin's theory of the origin of species.

A physical system in which randomness at the macroscopic level plays an important part is the diffusion of molecules in a fluid. Suppose, as shown in Figure L6.3.1, that a partition dividing a box into equal volumes is removed to allow diffusion into the originally empty region. Suppose that there is a total of N molecules. After n molecules have changed sides (L to R, or R to L), what proportion, $p_R(n)$, of the molecules are in the right-hand side? For example, Figure L6.3.1 has $N = 9$. Assume that only $n = 3$ molecules have passed through the partition. (This is only the least possibility consistent with the figure.) What is the probability for this arrangement?

There are two ways to approach this problem: First, we can set up an equation that relates the change in p_R to a change in n, then solve this equation analytically. Second, we can simulate the behavior of the gas by randomly transferring particles between left and right sides, but with a probability proportional to the numbers of particles in each side.

Analytic method

In the analytic method of investigating the approach to thermodynamic equilibrium, suppose that a total of n molecules have already exchanged sides. Now, if a relatively small number Δn are to change sides, then there will be a corresponding change in the proportion in the right side, $\Delta p_R(n)$. The gain to the right will be proportional to the number on the left, p_L, and the loss from the right will be proportional to the number on the right, p_R. This loss subtracts in the effect on Δp_R. Expressed algebraically,

$$\Delta p_R(n) = (p_L - p_R)\Delta n/N. \tag{L6.6}$$

▶ *Exercise* L6.3.1   Work through in detail the steps leading from the verbal arguments about how the proportions in R and L change to the algebraic result (L6.6).
□

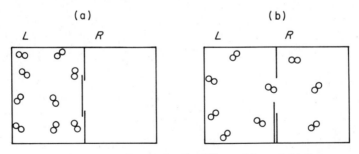

Figure L6.3.1   (a) Gas molecules (diatomic) are initially confined to the left (L) half of a container. In (b) they have been allowed to diffuse to the right (R) half.

To continue setting up the differential equation for the approach to thermodynamic equilibrium, we note that $p_L - p_R = 1 - 2p_R$, because the sum of the proportions in the two sides is unity. If we now take the limit as $\Delta n/N \to 0$, which requires that $N \gg 1$ since Δn can not be less than 1, we can approximate (L6.6) as a differential equation

$$D_n p_R(n) = [1 - 2p_R(n)]/N. \qquad (L6.7)$$

If we move the quantity $1 - 2p_R(n)$ to the left side, then we have a separable differential equation, as studied in A6.2. Both sides can be integrated with respect to n, to produce the indefinite integrals

$$\int dn\, D_n p_R/(1 - 2p_R) = \int dn/N. \qquad (L6.8)$$

The left side is readily integrated in terms of natural logarithms, to give

$$-\ln (1 - 2p_R)/2 = n/N + C, \qquad (L6.9)$$

where C is a constant of integration. With the initial condition that all the molecules are originally on the left side, then $p_R(0) = 0$, which fixes $C = 0$.

Solving explicitly for $p_R(n)$ in (L6.9) with $C = 0$, we obtain the analytic result for the proportion of molecules in the right side of the container after n interchanges of sides among the total of N molecules have occurred

$$p_R(n) = [1 - \exp(-2n/N)]/2. \qquad (L6.10)$$

Since $p_L + p_R = 1$, we obtain immediately for the number of molecules in the left side after n changes

$$p_L(n) = [1 + \exp(-2n/N)]/2. \qquad (L6.11)$$

These two algebraic expressions give the proportions of the molecules in each half of the container after n interchanges among the N molecules have occurred. By the time $2N$ molecules (here 18) have changed sides the proportions in the two sides are predicted to differ by less than 2%.

▶ *Exercise* L6.3.2   Make a graph of $p_R(n)$ against n/N for the range $n = 0$ to $n = 2N$. If you write a program to do this, it might be a good idea to use the plotting procedure from L4 to enable comparison with the results from the Monte-Carlo simulation that follows. □

These analytic developments enable us to calculate the proportion of the molecules in either side of the container. We now extend the discussion so that other interesting quantities may be calculated.

Entropy and the approach to equilibrium

An important concept in thermodynamics and in information theory is that of *entropy*. (The relationships between thermodynamics, statistical mechanics, and information theory are derived in a seminal book by Brillouin, while entropy in thermodynamics and statistical physics is covered in the book by Zemansky, Abbott, and Van Ness.) We define the entropy, S, as

$$S = \ln \, (\text{Probability}). \tag{L6.12}$$

In thermodynamics the entropy is usually defined to be Boltzmann's factor, k, times the entropy defined here. For us, the dimensional factor of k would be an unnecessary inconvenience.

▶ *Exercise* L6.3.3   Show that the total entropy of a system is the sum of the entropies of its parts. □

For the present system the entropy when n molecules have changed sides is $S(n)$, where

$$S(n) = \ln \, [p_L(n)] + \ln \, [p_R(n)]. \tag{L6.13}$$

An important principle, called the *second law of thermodynamics*, is that in a large system the entropy S always increases with time. In our system n, the number of molecules that have exchanged sides, will increase with time, so that the second law will be verified if $D_n S(n)$ is positive.

▶ *Exercise* L6.3.4   (Properties of entropy)
(a) Calculate the entropy $S(n)$ and its derivative with respect to n, by using equations (L6.10) and (L6.11). (This assumes that n can be approximated as a continuous variable.) Show that

$$D_n S(n) = (4/N)/[\, \exp \, (4n/N) - 1], \tag{L6.14}$$

which is always positive, so that the entropy indeed increases.
(b) Show that when relatively few molecules have exchanged sides ($n/N \ll 1$), a Maclaurin expansion (see A3.4) of the exponential factors can be made to give the approximation for the entropy

$$S(n) \approx \ln \, (n/N), \quad n/N \ll 1. \tag{L6.15}$$

(c) Show that if $n/N \gg 1$, so that relatively many exchanges have taken place, then the entropy tends to $-2 \ln 2 = -1.39$.
(d) On the basis of these results, sketch a graph of $S(n)$ against n/N. It would be practical to write a program to do the calculations and the plotting (using the procedure in L4), provided that the result implied in (b) that $S(n)$ diverges to $-\infty$ for

$n \to 0$, is accounted for, for example by starting the graph at $n/N = 1/e^4 = 0.018$.
□

Thus, in our model for the approach to thermodynamic equilibrium of a gas, and under the assumption that the change in the number of molecules is small compared with the total number of molecules, which allows the derivative approximation in (L6.14), the entropy of the system is predicted to increase. It will be interesting to see how well this holds in the simulation study in which N is not necessarily very large, so that $1/N$ (which is the smallest fractional change in n) is not necessarily very small.

Monte-Carlo method

The Monte-Carlo simulation model for the approach to thermal equilibrium uses random numbers to decide which molecules change sides. The Ith molecule is represented in a computer program by the array element MOL(I), where I ranges from 1 to the number of molecules, NMOL $= N$. Therefore, since even small volumes of gases at usual pressures contain zillions of molecules, this simulation is impractical for most real-world applications, except for the rarified reaches of outer space, which have a few molecules in each milliliter. However, it is very interesting to contrast the behavior of small and large ensembles.

Our Monte-Carlo simulation program proceeds as follows. The array element MOL(I) is to contain -1 if the molecule is on the left, and $+1$ if the molecule is on the right. Thus, if a molecule changes sides, we (or the computer) change signs. To decide which molecule changes sides on the nth interchange we compute I $=r_n*$NMOL, where r_n is a random number between 0 and 1, exclusive. (There is a chance that I will vanish; if so, use 1, then organize a search party.) Note that array locations will repeat because the (truncated) integral value of I repeats about once in NMOL tries. However, the modulus for the random-number generator, M, must be very much larger than NMOL in order that a periodic interchange of molecules is not initiated.

How does all this relate to the calculation of the proportion of molecules in the right side, p_R? After n interchanges, if we sum MOL(I) from I = 1 to NMOL, we will get twice the excess number of particles on the right. Therefore, we have that

$$p_R(n) = (1 + \text{sum/NMOL})/2. \qquad (L6.16)$$

▶ *Exercise* L6.3.5   Write out the detailed arguments that lead to the result (L6.16) for p_R. In particular, why does the sum give twice the excess, and why is there a factor of 2 in the last equation? □

With these preliminaries we can now outline the program in pseudocode:

Thermodynamic equilibrium simulation program

Input NMOL, NTRY, NSHOW and
   random-number generator values c, r_1, M
Initialize array MOL to -1 (all molecules on left)
Loop over tries N from 1 to NTRY
   Change sign of MOL(r_nNMOL $+1$)
   Generate next random number
   If NSHOW is a factor of N then
      Loop over I from 1 to NMOL summing MOL(I);
      $p_L = (1 + \text{sum}/\text{NMOL})/2$; $p_R = 1 - p_L$.
      Calculate the entropy $S = \ln(p_L) + \ln(p_R)$
      Compute analytic values of p_L, p_R and S
      Output N with Monte-Carlo and analytic values
      of p_L, p_R and S
End loop over N

Go for more data or options

Program end.

With this program outline for the Monte-Carlo simulation, you will be ready to research the approach to thermodynamic equilibrium.

▶ *Exercise L6.3.6*   (Program for approach to equilibrium)
(*a*) Write and debug a program, based on the preceding pseudocode, to make a Monte-Carlo simulation of the approach to equilibrium of a gas initially all in the left half of a container. Note that debugging is not trivial because of the pseudo-randomness of the choice of molecules. Build into your program options that allow you to print out step-by-step values of all variables; for example, set NSHOW $= 1$.
(*b*) Make a simulation with say 100 molecules and make 200 tries. Plot the simulation and analytic values of $p_R(n)$ against n/N. How good is the agreement? Are discrepancies consistent with the $\sqrt{n}$ variation expected for Monte-Carlo simulations?
(*c*) Compare the entropy obtained from the simulation results and from the analytic formulas for the probabilities. Use the same data as in (*b*). Does the entropy *always* increase? Which is the more correct formulation, simulation or analytic?
(*d*) Run your program with only 10 molecules and still 200 tries. Why is the disagreement between simulation and analytic approximation more severe than in (*c*)? Which is the less realistic approximation?
(*e*) Suppose that you are sitting in a room with calm air at normal pressure. Is there much chance that you will suddenly be asphyxiated because most of the air molecules leave the volume around your head? Relate your argument to your preceding results. □

In this section we have examined a real-world situation modeled by a Monte-Carlo simulation. The next simulation example models a much smaller system, namely the decay of a sample of radioactive nuclei.

L6.4 {2}   Simulation of nuclear radioactivity

In nuclear radioactivity the rate of decay of a sample that is not too "hot" is often sufficiently small that discrete decays can be detected. Therefore the Monte-Carlo simulation may be much more realistic than the continuum approximation made in an analytic model, in which the activity of the source is predicted to decay smoothly and exponentially with time. However, we begin with an analytic model for calculating the decay rate, since it gives insight into setting up the Monte-Carlo simulation.

Analytic formula for radioactivity

Suppose that we have a radioactive sample of atoms which are not being created but only can decay. Experimentally it was found in the early twentieth century that the fraction $\Delta P/P$ of the decays in a small time interval Δt is proportional to Δt. (This is discussed in the physics text by Krane and in the book on radioactivity measurements by Mann, Ayres, and Garfinkel.) Thus

$$\Delta P/P = -\lambda \Delta t. \tag{L6.17}$$

Here the radioactive decay constant λ is the (positive) constant of proportionality. The units of λ are reciprocal time. In the limit that $\Delta t \to 0$, and assuming that some nuclei still decay in this time interval, this equation can be approximated as a differential equation

$$D_t P(t)/P(t) = -\lambda. \tag{L6.18}$$

If both sides are integrated with respect to t, then we readily find that

$$\ln P(t) = -\lambda t + C, \tag{L6.19}$$

where C is a constant of integration.

▶ *Exercise* L6.4.1   (Differential equation for radioactivity)
(*a*) Derive the radioactivity differential equation, then integrate it to produce (L6.19).
(*b*) Verify (L6.19) by differentiating it and recovering (L6.18).
(*c*) Show that the constant of proportionality in (L6.19) is $C = \ln[P(0)]$, where $P(0)$ is the number of nuclei present when the counting of the radioactivity began. □

On combining (L6.19) and the value obtained in Exercise L6.4.1 for C, the continuum approximation to radioactive decay predicts that after a time t the number of nuclei that have not yet decayed is $P(t)$ given by

$$P(t) = P(0) \exp(-\lambda t). \tag{L6.20}$$

We can write this alternatively as the proportion of the original nuclei that survive to time t is $p(t)$, given by

$$p(t) = \exp(-\lambda t). \tag{L6.21}$$

▶ *Exercise* L6.4.2   Suppose that the radioactive material is enclosed in a box from which neither the parent nor daughter atoms can escape. If the decay products came into equilibrium with the original atoms, then would we not get the same formula for $p(t)$ as for $p_L(n)$ in the approach to thermodynamic equilibrium? Why does this not actually happen? □

Monte-Carlo simulation

Our Monte-Carlo simulation of radioactive decay uses random numbers to decide whether or not a nucleus is to decay. Specifically, the program uses random numbers in the range 0 to 1 in order to decide whether a nucleus that has not yet decayed should be given a chance to decay. If the random number is less than λ, then the nucleus decays, otherwise it is left alone. This sampling of random numbers is to be done for each nucleus, of which there are NUCL. After each NSHOW samples the results are to be output. The pseudocode outline of the program is as follows:

Radioactive decay Monte-Carlo simulation program

Input NUCL, NTRY, NSHOW, LAMBDA and random number
    controls c, r_1, M.
Loop over nuclei I from 1 to NUCL
    Initialize each PARENT(I) to 1
End loop over nuclei
Loop over N (trial number) from 1 to NTRY
    Loop over all nuclei I from 1 to NUCL
      If PARENT(I) = 1 then
        Generate a random number in 0 to 1
        If random number < LAMBDA set PARENT(I) = 0
        and NDECAY = NDECAY + 1
    End loop over nuclei
    If NSHOW factors N then output proportion not yet decayed,
      1 − NDECAY/NUCL

End loop over trials

Go for more data or options

Program end.

► *Exercise L6.4.3* (Simulating radioactivity)
(*a*) Write and debug a program that simulates radioactive decay according to the scheme outlined in the preceding pseudocode. Use optional output statements so that you can debug your program by stepping through it, for example with NSHOW = 1.
(*b*) Consider a small sample size, NUCL, and study its decay in detail, graphing $p(N)$ and comparing it with the exponential decay formula (L6.21). Are the results significantly different, according to the $\sqrt{N}$ argument derived as (L6.4)?
(*c*) Now run a large sample size of say 10,000 nuclei (if this is allowed in your computer system). Have the relative fluctuations in the decay rate decreased? How different are the results from the analytic formula? Are the differences significant? □

The Monte-Carlo simulations in these two sections have been detailed simulations of actual phenomena, either molecules moving between two sides of a container, or atomic nuclei undergoing radioactive disintegration. At the microscopic level the simulations are more realistic than their analytic approximations, because they use discrete events. However, note that we have not provided dynamical models of the phenomena, but only descriptions. What forces are there to make molecules change sides, and why does a nuclear system undergo radioactive decay? These are questions that researchers in molecular dynamics and nuclear physics puzzle over. For an introduction read Zemansky, Abbott and Van Ness on thermodynamics, and Kane or Mann, Ayres and Garfinkel on radioactivity.

Other Monte-Carlo simulations

There are many other interesting and instructive examples on simulations by random-number techniques applied to the sciences. For example, the textbook by Bennett offers several examples that can be programmed in BASIC. There are simulation electronic games with programs for microcomputers in the book by Engel.

Further computer applications to thermodynamics and statistical mechanics are given in the books by Ehrlich and by Merrill, and in the article by Sauer. For more advanced students there are textbooks, programs, and articles by Nakamura (particle transport and heat transfer), by Edelson (chemical kinetics simulations), by Wetherill (planet formation), and by Boardman (electrons in semiconductors).

Random references on Monte-Carlo methods

Bennett, W. R., *Scientific and Engineering Problem Solving with the Computer*, Prentice-Hall, Englewood Cliffs, N.J., 1976.

Boardman, A. D. (ed.), *Physics Programs*, Wiley, New York, 1980.

Brillouin, L., *Science and Information Theory*, Academic Press, New York, 1962.

Dudewicz, E. J. and T. G. Ralley, *The Handbook of Random Number Generation and Testing with TESTRAND Computer Code*, American Sciences Press, Columbus, Ohio, 1981.

Edelson, D., "Computer Simulation in Chemical Kinetics," *Science*, **214**, 981 (1981).

Ehrlich, R., *Physics and Computers*, Houghton Mifflin, Boston, 1973.

Engel, C. W., *Stimulating Simulations*, Hayden Book Co., Rochelle Park N.J., 1977.

Frigerio, N. A. and N. A. Clark, "A Random Number Set for Monte-Carlo Computations," *Transactions of the American Nuclear Society*, **22**, 238 (1975).

Gamow, G., *One Two Three ... Infinity*, Viking Press, New York, 1947, Chap. 8.

Hall, A., "An Experimental Determination of π," *The Messenger of Mathematics*, **2**, 113 (1873).

Knuth, D. A., *The Art of Computer Programming*, Vol.2, Addison-Wesley, Reading, Mass., (1969) Chap.3.

Krane, K. S., *Modern Physics*, Wiley, New York, 1983.

Levison, M., R. G. Ward, and J. W. Webb, *The Settlement of Polynesia*: *A Computer Simulation*, University of Minnesota Press, Minneapolis, 1973.

Mann, W. B., R. L. Ayres, and S. B. Garfinkel, *Radioactivity and Its Measurement*, Pergamon Press, Oxford, 1980.

Merrill, J. R., *Using Computers in Physics*, Houghton Mifflin, Boston, 1976.

Mishima, N., T. Y. Petrovsky, H. Minava, and S. Goto, "Model Experiment of Two-Dimensional Brownian Motion by Microcomputer," *American Journal of Physics*, **48**, 1050 (1980).

Nakamura, S., *Computational Methods in Engineering and Science*, Wiley, New York, 1977.

Sauer, G., "Teaching Statistical Mechanics; A Simulation Approach," *American Journal of Physics*, **49**, 13 (1981).

Wetherill, G. W., "The Formation of the Earth from Planetesimals," *Scientific American*, June 1981, p.163.

Zemansky, M. W., M. M. Abbott, and H. C. Van Ness, *Basic Engineering Thermodynamics*, McGraw-Hill, New York, 1975.

L7   SPLINE FITTING AND INTERPOLATION

Have you ever watched an experienced technical artist drawing a nice, smooth curve using a drafting spline? You now know from section A4.6 that this procedure can be automated. This laboratory provides a practical introduction to automatic calculation of spline curves. Practical results are emphasized because spline analysis involves a sequence of operations expressed as an algorithm, rather than a small set of readily comprehensible formulas for the spline coefficients. Thus, although you may have a good intuitive idea of spline fitting from drafting practice, numerical splines are probably still quite mysterious to you.

The theory of curve fitting by splines is detailed in A4.6, where an algorithm for computing cubic spline coefficients for equally spaced knots is derived. Spline fitting in numerical analysis and applied science is not widely understood, despite its inherent simplicity. Therefore in L7.1 I have provided sample programs for spline fitting. Programs in both Pascal and Fortran are given. With these programs, and modifications or enhancements to them, you will be able to begin exploring spline analysis. Several computing exercises on cubic splines are suggested in L7.2. These exercises involve investigating the effects of various endpoint conditions, interpolation, differentiation, and integration. Further study of spline fitting is suggested by the references at the end of the chapter. Extensive programs, coded in Fortran, are presented in De Boor's book.

L7.1   {2}   Sample programs for spline fitting

To encourage you to understand and use spline fitting, I have included programs for spline fitting and interpolation in two common scientific computing languages, Pascal and Fortran. The two programs are coded similarly, thereby compromising a little on the elegance of the Pascal version. If you are an *aficionado* of one of these languages, why not at least become *curioso* about the other and read it over?

To show the common origin of both programs, the structure of the procedures is first indicated. If you code the spline algorithms in another language, such as BASIC, you will probably find the following pseudocode very helpful.

Structure of the spline procedures

In A4.6 we derived an algorithm for cubic spline fitting and interpolation of data with equally spaced knots (independent variables). The five sections of this algorithm are indicated in the following by the same numbers. The formulas within each section are obtained from A4.6. The procedures divide naturally into two program segments, SPLFIT for determining the fitting coefficients, and SPLINT for interpolating values and derivatives.

All the looping over the knot index k may appear unnecessary. Why not have just a single loop? The main justifications for multiple loops are simplicity of coding and clarity. If you try a different loop structure, you will probably have a much more complicated program, without any real gain in execution speed.

L7 Spline fitting procedures

SPLFIT: Cubic spline fit with equal knot spacing
1) Spline coefficients
    Loop over k increasing to compute a_k

2) Differences;
    First differences
        Loop over k increasing for e_k
    Second differences
        Loop over k increasing for d_k
        Loop over k increasing for b_k

3) Second derivatives; spline estimates
    Loop over k decreasing for s''_k

4) First and third derivatives; spline estimates
    Loop over k increasing for first derivative
    Loop over k increasing for third derivatives

5) Interpolation
    Input an x value in range x_1 to x_n
    Locate nearest smaller interpolation index, i
    Compute distance above index, e
    Interpolate values from Taylor expansion about x_i
    Derivatives
        Estimate first, second and third derivatives
            by interpolation
    Go for next x value

Program end.

The testing of the computer programs is discussed in the subsections containing the coding. The question of the effects of numerical noise (such as truncation and round-off errors) on the spline calculations is considered in the book by Vandergraft. One might suspect that subtractive cancellation, as discussed in A4.2, could become important if there is a large number of data points, n, because there are several loop cycles, especially in 2) and 3), where subtraction of adjacent values is performed.

Spline fitting from Pascal

Coding of the Pascal procedures for spline fits follows directly from the pseudocode outline in the preceeding subsection. The main procedure, SPLCHK, is used for input of the spline variables, to invoke the spline fitting routine, SPLFIT, then to allow interpolation to x values using procedure SPLINT. It should be emphasized that the purpose of the main routines as written is only to provide checking input to the program, and not to serve as a general procedure for use in extensive data analysis. For the latter requirement, you should tailor the program to your own particular needs and type of input devices, for example, batch input or interactive input, error-free or noisy data values. Thus, the values of x that are input should lie within the range of values for which the spline fit has been made, but it is up to you to provide the programming to make this checking if you consider it necessary for your applications.

The code presented here executes correctly on an Apple II + microcomputer and uses only Pascal statements compatible with UCSD Pascal.

```
PROGRAM SPLCHK;
(* Cubic spline fits checking; Pascal *)
VAR N,IK:INTEGER;
    XMIN,XMAX,HSTEP,X,E,SX,SD1X,SD2X,SD3X:REAL;
    YLIST,SD1,SD2,SD3:ARRAY[1..100] OF REAL;
PROCEDURE SPLFIT;
(* Fit of YLIST on X at N equal steps HSTEP *)
VAR K:INTEGER;
    HFACT,HD2,NM1:REAL;
    AS,ES,DS,BS:ARRAY[1..100] OF REAL;
BEGIN
 HFACT:=6/(HSTEP*HSTEP); HD2:=HSTEP/2;
 NM1:=N-1;
(* 1) Spline coefficients *)
 AS[2]:=4;
 FOR K:=3 TO NM1 DO AS[K]:=4-1/AS[K-1];
(* 2) Differences *)
 (* First differences *)
 FOR K:=2 TO N DO ES[K]:=YLIST[K]-YLIST[K-1];
 (* Second differences *)
 FOR K:=2 TO NM1 DO DS[K]:=HFACT*(ES[K+1]-ES[K]);
 DS[2]:=DS[2]-SD2[1]; DS[NM1]:=DS[NM1]-SD2[N];
 (* b coefficients *)
 BS[2]:=DS[2];
 FOR K:=3 TO NM1 DO BS[K]:=DS[K]-BS[K-1]/AS[K-1];
(* 3) Second derivatives; spline estimates *)
 (* Endpoint values SD2[1] and SD2[N] input *)
 SD2[NM1]:=BS[NM1]/AS[NM1];
 FOR K:=N-2 DOWNTO 2 DO
 SD2[K]:=(BS[K]-SD2[K+1])/AS[K];
```

```
(* 4) First and third derivatives *)
 (* First derivatives *)
 SD1[1]:=ES[2]/HSTEP-(SD2[1]/3+SD2[2]/6)*HSTEP;
 FOR K:=2 TO N DO
 SD1[K]:=SD1[K-1]+(SD2[K-1]+SD2[K])*HD2;
 (* Third derivatives *)
 FOR K:=1 TO NM1 DO
 SD3[K]:=(SD2[K+1]-SD2[K])/HSTEP;
 SD3[N]:=SD3[NM1];
END; (* SPLFIT *)
PROCEDURE SPLINT;
(* Spline interpolation; derivatives in SPLFIT *)
VAR I:INTEGER;
    E:REAL;
BEGIN
 I:=TRUNC((X-XMIN)/HSTEP)+1; (* Interpolation *)
 E:=X-XMIN-(I-1)*HSTEP; (* distance from I *)
 (* Interpolated SX: Derivatives; first SD1X,
                      second SD2X, third SD3X *)
 SX:=YLIST[I]+(SD1[I]+(SD2[I]/2+SD3[I]*E/6)*E)*E;
 SD1X:=SD1[I]+(SD2[I]+SD3[I]*E/2)*E;
 SD2X:=SD2[I]+SD3[I]*E; SD3X:=SD3[I];
END; (* SPLINT *)
BEGIN
 WRITELN('Spline Checking program; Pascal');
 WRITE('Number of data, N?');READLN(N);
 WRITELN('Input ',N:2,' Y values:');
 FOR IK:=1 TO N DO READLN(YLIST[IK]);
 WRITELN('Second derivatives at K=1 & K=N?');
 READLN(SD2[1],SD2[N]);
 WRITELN('Minimum and maximum X values?');
 READLN(XMIN,XMAX);
 HSTEP:=(XMAX-XMIN)/(N-1); (* uniform step *)
 SPLFIT; (* Spline fit of N values in YLIST *)
 WHILE NOT(EOF) DO
 BEGIN
  WRITE('X value?');READLN(X);
  SPLINT; (* Spline interpolation to point X *)
  WRITELN('S(X),SD1(X),SD2(X),SD3(X);',
          SX,SD1X,SD2X,SD3X);
 END;
END.
```

This Pascal code was tested by feeding in the exact values for $y(x) = x^3$, which should be described exactly by the cubic spline if the correct endpoint second derivatives, $s''(x_1)$ and $s''(x_n)$, are input. This is easily done because we know that $y'(x) = 3x^2$ and $y''(x) = 6x$. Actually, I initially had trouble debugging with the cubic, so I went back to a linear function, debugged this, moved up to a quadratic, then finally to the cubic. By this means I was able to isolate the parts of the program that were in error.

▶ *Exercise* L7.1.1   Explain the debugging strategy by which the cubic spline procedure is efficiently debugged by testing with linear, quadratic, then cubic, functions.
   □

Spline fitting from Fortran

The coding of the Fortran version of the cubic spline fitting algorithm follows the structure of the Pascal procedure in the previous subsection. I transcribed the Pascal statements into Fortran and made the Pascal global variables into arguments passed between subroutines through the argument lists in the Fortran calls. By this means it took only three compilations to obtain the present Fortran program.

The Fortran cubic spline fitting and interpolation program has been tested on the Apple II + microcomputer using Apple Fortran, and it has been tested on a macrocomputer using WATFIV, as described in the book by Dyck, Lawson, and Smith. The Fortran and Pascal test outputs with a cubic func-tion input are essentially identical.

```
      PROGRAM SPLCHK
C   Cubic spline fits checking; Fortran
      INTEGER N,IK
      REAL XMIN,XMAX,HSTEP,X,E,SX,SD1X,SD2X,SD3X,
     1      YLIST(100),SD1(100),SD2(100),SD3(100)
C   A * denotes terminal output or input
      WRITE(*,12)
   12 FORMAT(' Spline Checking program; Fortran')
      WRITE(*,14)
   14 FORMAT(' Number of data, N?')
      READ(*,16) N
   16 FORMAT(I2)
      WRITE(*,18) N
   18 FORMAT(' Input ',I2,' Y values:')
      READ(*,20) (YLIST(IK),IK=1,N)
   20 FORMAT(F5.2)
      WRITE(*,22)
   22 FORMAT(' Second derivatives at K=1 & K=N?')
      READ(*,24) SD2(1),SD2(N)
   24 FORMAT(2F5.2)
```

```
      WRITE(*,26)
   26 FORMAT(' Minimum and maximum X values?')
      READ(*,28) XMIN,XMAX
   28 FORMAT(2F5.2)
C  Uniform step size
      HSTEP = (XMAX-XMIN)/(N-1)
C  Cubic spline fit of N values in YLIST
      CALL SPLFIT(N,HSTEP,YLIST,SD1,SD2,SD3)
C  Cubic spline interpolation to point X
   30 WRITE(*,31)
   31 FORMAT(' X value?');
      READ(*,32) X
   32 FORMAT(F5.2)
      CALL SPLINT(X,XMIN,HSTEP,YLIST,SD1,SD2,SD3,
     1            SX,SD1X,SD2X,SD3X)
      WRITE(*,34) SX,SD1X,SD2X,SD3X
   34 FORMAT(' S(X),SD1(X),SD2(X),SD3(X);',
     1         4(1X,F6.2))
      GO TO 30
      END
      SUBROUTINE SPLFIT(N,HSTEP,YLIST,SD1,SD2,SD3)
C  Fit of YLIST on X at N equal steps HSTEP
      INTEGER K,NM1,NM2,NMK
      REAL HFACT,HD2,
     1     AS(100),ES(100),DS(100),BS(100),
     2     YLIST(100),SD1(100),SD2(100),SD3(100)
      HFACT = 6/(HSTEP*HSTEP)
      HD2 = HSTEP/2
      NM1 = N-1
      NM2 = N-2
C  1)  Spline coefficients
      AS(2) = 4
      DO 100 K = 3,NM1
      AS(K) = 4-1/AS(K-1)
  100 CONTINUE
C  2)  Differences
C    First differences
      DO 200 K = 2,N
      ES(K) = YLIST(K)-YLIST(K-1)
  200 CONTINUE
C    Second differences
      DO 300 K = 2,NM1
      DS(K) = HFACT*(ES(K+1)-ES(K))
  300 CONTINUE
      DS(2) = DS(2)-SD2(1)
```

```
               DS(NM1) = DS(NM1)-SD2(N)
C     b coefficients
               BS(2) = DS(2)
             DO 400 K = 3,NM1
               BS(K) = DS(K)-BS(K-1)/AS(K-1)
       400 CONTINUE
C  3) Second derivatives; spline estimates
C     Endpoint values SD2(1) and SD2(N) input
               SD2(NM1) = BS(NM1)/AS(NM1)
C  Downwards iteration
             DO 500 K = 2,NM2
               NMK = N-K
               SD2(NMK) = (BS(NMK)-SD2(NMK+1))/AS(NMK)
       500 CONTINUE
C  4)  First and third derivatives
C     First derivatives
               SD1(1) = ES(2)/HSTEP-
             1          (SD2(1)/3+SD2(2)/6)*HSTEP
             DO 600 K = 2,N
               SD1(K) = SD1(K-1)+(SD2(K-1)+SD2(K))*HD2
       600 CONTINUE
C     Third derivatives
             DO 700 K = 1,NM1
               SD3(K) = (SD2(K+1)-SD2(K))/HSTEP
       700 CONTINUE
               SD3(N) = SD3(NM1)
             RETURN
             END
             SUBROUTINE SPLINT(X,XMIN,HSTEP,YLIST,
             1                    SD1,SD2,SD3,
             2                    SX,SD1X,SD2X,SD3X)
C  Spline interpolation; derivatives in SPLFIT
             INTEGER I
             REAL X,XMIN,HSTEP,E,
             1      YLIST(100),SD1(100),SD2(100),SD3(100),
             2      SX,SD1X,SD2X,SD3X
C  Interpolation index
             I = (X-XMIN)/HSTEP+1.1
C  Distance from Ith point
             E = X-XMIN-(I-1)*HSTEP
C  Interpolated SX: Derivatives; first SD1X,
C                   second S2DX, third SD3X
             SX = YLIST(I)
             1      (SD1(I)+(SD2(I)/2+SD3(I)*E/6)*E)*E
```

```
SD1X = SD1(I)+(SD2(I)+SD3(I)*E/2)*E
SD2X = SD2(I)+SD3(I)*E
SD3X = SD3(I)
RETURN
END
```

The next section gives you several opportunities to use and modify these cubic spline programs.

L7.2 {2}   Examples using splines

The examples in this section first check out the sensitivity of the spline fits to the values of the second derivatives at the endpoints, the boundary conditions on the spline analysis. I think that you will discover that the so-called "natural spline" conditions, which have both endpoint second derivatives zero, do not always give a good representation of the curve.

The other exercises give you a chance to see how well spline fits suffice for interpolation, for first and second derivatives, and for integration.

Boundary condition and interpolation exercises

▶ *Exercise* L7.2.1   (Natural and other spline boundary conditions)
This series of exercises uses the program SPLFIT from section L7.1 to examine the sensitivity of spline fits to the endpoint conditions on the second derivatives. Since we are studying *cubic* splines, it is a good idea to use as test data a cubic, $y(x) = ax^3 + bx^2 + cx + d$, in which the coefficients and range of x values (equally spaced) are chosen by you. Then you can compute the derivatives that you expect exactly and use these for comparison with the spline results.
(*a*) In SPLFIT, try the natural spline boundary conditions on the endpoint second derivatives, SD2(1) = 0, SD2(N) = 0. How well do the interpolated values and estimated derivatives from SPLINT agree with exact values? Do any discrepancies decrease as you move to points k further away from the ends? Are the errors least severe for the interpolated function and the low-order derivatives?
(*b*) Try for the second-derivative boundary conditions SD2(1) and SD2(N) the exact value at the midpoint of the x range. Does this generally improve the fits over those obtained using the natural spline in (*a*)?
(*c*) Here is an experiment on spline fitting. Begin the spline analysis with the "natural" boundary conditions. Use this to predict the values of the second derivatives at the endpoints. Repeat the spline analysis with these as boundary conditions. If you do this iteratively does the agreement with exact results steadily improve? Or, does it converge to a consistent but incorrect value? □

If you have completed this exercise, you will appreciate the importance of the endpoint conditions to spline fitting. Therefore, you can choose appropriately accurate conditions in the following exercises.

► *Exercise L7.2.2*   (Interpolation using cubic splines)
This exercise emphasizes use of the interpolation procedure SPLINT after the spline-fitting program SPLFIT coded in section L7.1. It is probably worthwhile to improve the driver program to allow more flexible control of input options.
(*a*) Verify the correctness of the cubic-spline interpolation routine SPLINT by using as data a single cubic polynomial and the exact second derivatives at the endpoints. Interpolation to values at, say, midway between the knots should be exact.
(*b*) Investigate the practicality of using cubic splines to reduce the size of function tabulations that must be stored. Note that a cubic spline fit through n knots generates $4n$ entries, because derivatives at the knots are required. Therefore, unless the table can be reduced in size by at least a factor of four, there is no gain in storage, but probably a loss in accuracy and speed. However, if the original function is very expensive to generate, even this may be a saving.
   Test the interpolation procedure on, for example, the function $y(x) = \cos(x)$ for $0 \leqslant x \leqslant \pi/2$, using the correct endpoint derivatives, SD2(1) = -1, SD2(N) = 0. Calculate, and comment on, the quality of the interpolated values.
(*c*) The interpolation formula for the function value SX requires eight arithmetic operations. Compare the execution speed of the SPLINT routine (computing only SX) with the evaluation of the original function, for example, the cosine. To get fair accuracy in your speed estimates, you may need to evaluate of the order of 100 such expressions in both methods.
(*d*) In computing laboratory L10 on the analysis of resonance line widths, there are given data on the microwave absorption inversion resonance in ammonia (section L10.3). Using the 11 data points given there, make a natural cubic spline fit to the resonance data. Estimate the resonance frequency f_0 by varying the frequency at which the interpolation is made until the interpolated slope SD1 is essentially zero. Does your estimate of the resonance frequency agree with that quoted in L10.3, $f_0 = 23.87$ GHz?
(*e*) Refer to a handbook of mathematical functions, such as that edited by Abramowitz and Stegun, and choose a complicated looking function. How well can you interpolate among the table entries using cubic splines with a reasonable number of entries? □

Exercises on spline derivatives

The next set of exercises on cubic splines relates to the estimation of derivatives. This sensitive subject, considered from the viewpoint of Taylor expansions in A4.3 and L3, can also be approached from the spline analysis.

► *Exercise L7.2.3*   (Derivatives from cubic splines)
   In this exercise you are invited to compare the first and second interpolated

derivative estimates SD1 and SD2 with exact values. Therefore, simple known functions, whose derivative formulas you can easily include within SPLCHK for comparison purposes, are suggested.

(*a*) Consider the cosine function in the range 0 to π. Compare the spline estimates for various endpoint conditions of the first and second derivatives with the analytic values $-\sin(x)$ and $-\cos(x)$.

(*b*) Compare the spline estimates with the results obtained by the methods used for numerical differentiation in L3.

(*c*) For the function $y(x) = \sqrt{x}$ and the range, say $0 \leqslant x \leqslant 10$ in steps of unity, are the derivatives at midpoints of steps reproduced by the natural cubic spline? □

Spline integration exercises

Finally, you can exploit the power of splines to show their effectiveness for computing integrals of functions and data.

▶ *Exercise L7.2.4*   (Integration by cubic splines)

This exercise consists of algebraic and numerical parts that develop and test spline integration techniques.

(*a*) Consider the indefinite integral, but confined to x in the ith region of the spline fit, $i = 1,\ldots,n - 1$. Let $I_i(x) = \int dx' \, y(x')$. Show that use of the cubic spline approximation to $y(x)$, namely $s(x)$, results in

$$I_i(x) \approx (y_i + (s'_i/2 + (s''_i/6 + s'''_i x/24)x)x)x. \tag{L7.1}$$

Thence use additivity of integrals to show that for i and j in the range 1 to $n - 1$ the cubic spline estimate of the integral from x_i to x_j is

$$\int dx \, y(x) \approx \Sigma \, [I_k(x_{k+1}) - I_k(x_k)]. \tag{L7.2}$$

Here the summation is over the range $k = i$ to $k = j$. Note that the formula assumes that x_i and x_j coincide with knots at the lower and upper ranges of the integral. It is straightforward to remove this restriction by adding in integrals from portions of an x interval by using (L7.1).

(*b*) Modify the cubic spline program SPLCHK by adding a procedure, say SPLICE, that binds together the regions and performs the indicated summation. Debug SPLICE using functions whose integrals are well known and that can be readily incorporated into your program for checking purposes.

(*c*) Test the accuracy of the cubic spline integration formulas by integrating the cosine function form 0 to π, then from 0 to 2π. In making the spline fit, you might explore the effects of varying the endpoint conditions, as well as the number of knots. □

From the foregoing exercises you will gained practice and insight into computing with spline fits. You will be able to make extensive use of the Pascal or Fortran procedures SPLFIT in a variety of computing applications.

References on spline fitting

Abramowitz, M., and I. A. Stegun, *Handbook of Mathematical Functions*, Dover, New York, 1964.

De Boor, C., *A Practical Guide to Splines*, Springer-Verlag, New York, 1978.

Dyck, V. A., J. D. Lawson, and J. A. Smith, *Introduction to Computing*, Reston Publishing, Reston, Vir., 1979.

Vandergraft, J. S., *Introduction to Numerical Computations*, Academic Press, New York, 1978.

L8   LEAST-SQUARES ANALYSIS OF DATA

How do you fit a smooth curve through data that are known to be imprecise? In chapter A4 on numerical derivatives, integrals, and curve fitting we introduced the principle of the least-squares fit. In a least-squares analysis of data we choose a form for the curve through the data, then adjust the parameters determining the curve in such a way that the sum of the squares of the deviations of the fitted variable from the data is minimized. The choice of the least-squares criterion was discussed in A4.7, where we showed that if the fitting parameters appear linearly in the problem (linear least squares), then they can be determined by solving a set of linear equations.

In particular, if the relation between the data and the independent variable can be transformed to linear form, then we have a straight-line least-squares analysis. The intercept and slope for such a straight line can be determined from the data by algebraic formulas. The numerical stability of these formulas was investigated in Exercise A4.7.5 and in Problem [A4.10].

In this computing laboratory you can explore least-squares analysis of data as it applies to fitting straight lines. What if both the independent variable, x, and the dependent variable, y, are imprecise? We pose this problem and answer it (to the author's satisfaction) in section L8.1. In L8.2 we have sample computer programs in both Pascal and Fortran for the fitting of data by straight-line weighted least-squares, allowing for errors in both x and y. You can use these programs as starting points for stimulating exercises in the analysis of several interesting sets of data in both pure and applied science. These exercises are presented in section L8.3. The programs can also be used in the analysis of resonance line shapes by Fourier transform techniques, which is the topic of L10. A reading list, giving the sources of data used in our analyses and suggestions for further investigation of least squares, concludes the chapter.

L8.1 {2}   Straight-line fits with errors in both variables

In many experimental situations one of the variables can be adjusted, by a switch setting, a computer command, or a skilled experimenter, so that its value is relatively well determined. It is then called the independent variable, which we labeled x in chapter A4. Another variable in the experiment that changes as x changes is called the dependent variable, which we called y. It is usually assumed in least-squares analyses that the uncertainty or error in y is much greater than that in x. The situation is summarized in Figure L8.1.1.

But, neglect of the error in x compared with that in y is often unjustified. For example, both x and y may be variables over which we do not have much control, as in the visual readouts from two meters with comparable scale accuracies. Furthermore, we may not be sure whether it is more meaningful to plot y

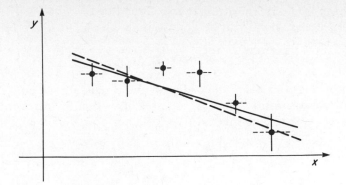

Figure L8.1.1   Straight-line least-squares fit: solid line fit
ignores errors in ordinates: dashed line fit includes equal
errors in ordinate and abscissa (as indicated).

against x, or x against y. In this section we derive a formula that, under
realistic assumptions about the relative errors in x and y, allows a simple calculation of the intercept and slope in a straight-line weighted least-squares fit,
without requiring significantly more computation than the usual case in which
the error in x is ignored. Prescriptions for calculating cases of greater generality, such as least-squares polynomial fits, are referenced in the article by
Pasachoff.

Straight-line least squares

To derive the straight-line least-squares formulas for the slope and intercept,
it is easier to begin at the definition of χ^2 from section A4.7 with the straight-
line assumption made explicitly

$$\chi^2 = \Sigma \left[y_k - (a_1 + a_2 x_k) \right]^2 \omega_k, \qquad (L8.1)$$

in which the sum is over k from 1 to n, the number of data points. We now
assume that the weight, ω_k, to be assigned to the point k is given in terms of
the error e_k at that point by $\omega_k = 1/e_k^2$, an assumption justified by the statistical theory of error analysis. The error has two origins: The first is from
indeterminacy in y; call this error $\sigma_k(y)$. The second error is from the uncer-
yainty in x, say $\sigma_k(x)$. This variation in x causes an uncertainty $a_2\sigma_k(x)$ in the
value of y, even if y were measurable without error. If the sources of the x and y
errors are independent, then the total error can be obtained from

$$e_k^2 = \sigma_k^2(y) + a_2^2 \sigma_k^2(x) \qquad (L8.2)$$

$$\mu = \sigma_k^2(x)/\sigma_k^2(y), \qquad (L8.3)$$

A tractable solution for the straight-line fit can be obtained only if the ratio of x to y errors at each point can be approximated by a constant. Define

$$\mu = \sigma_k^2(x)/\sigma_k^2(y), \tag{L8.3}$$

For example, in Figure L8.1.1 the ordinate and abscissa have equal errors, so $\mu = 1$. The usual assumption in least-squares fits is that the error in the x variable is relatively negligible, that is, $\mu = 0$. The total weight at the kth data point is given by

$$\omega_k = w_k(1 + a_2^2\mu), \tag{L8.4}$$

where $w_k = 1/\sigma_k^2(y)$ is the weight associated with the error in y. Notice that when we vary a_2 in order to minimize χ^2, we have to allow for its variation in the weight as well as in the difference between datum and straight-line fit. However, some symmetry has been gained by this, since the x and y variables are treated on an equal footing, as the subsequent analysis shows.

It is straightforward but uninspiring algebra to differentiate χ^2 with respect to a_1 and then with respect to a_2 and to derive the straight-line least squares formulas.

Least-squares formulas

In the following let

$$S_w = \Sigma\, w_k, \quad S_x = \Sigma\, x_k w_k, \quad S_y = \Sigma\, y_k w_k, \tag{L8.5}$$

in which the sums are over k from 1 to n. On setting the derivative of χ^2, obtained from (L8.1) combined with (L8.4), with respect to a_1 to zero, you will readily find that the intercept on the x axis is given by

$$a_1 = y_{av} - a_2 x_{av}, \tag{L8.6}$$

where the weighted averages are given by

$$x_{av} = S_x/S_w, \quad y_{av} = S_y/S_w. \tag{L8.7}$$

This result for a_1 is not immediately usable, because it contains the unknown slope a_2. However, it allows us to set in the calculation of χ^2

$$y_k - a_1 - a_2 x_k = y_k - y_{av} - a_2(x_k - x_{av}). \tag{L8.8}$$

This also has the numerical advantage of minimizing subtractive cancellation, as you perhaps discovered in Problem [A4.10].

▶ *Exercise* L8.1.1 Derive the equation for the slope a_2 by substituting (L8.8) in the formula for the derivative of χ^2 with respect to a_2 to show that

$$a_2\mu S_{xy} + (S_{xx} - \mu S_{yy}) - (1/a_2)S_{xy} = 0, \qquad (L8.9)$$

where $S_{xy} = \Sigma (x_k - x_{av})(y_k - y_{av})w_k$, and similarly for S_{xx} and S_{yy}. Note that equation (L8.9) for a_2 is completely symmetric between x and y, remembering that the slope of x on y is $1/a_2$ if that for y on x is a_2, as here, and that $\mu \rightarrow 1/\mu$ under interchange of x and y. □

The quadratic equation for the slope a_2, (L8.9), has the standard solution

$$a_2 = \{ - (S_{xx} - \mu S_{yy}) + \sqrt{[(S_{xx} - \mu S_{yy})^2 + 4\mu S_{xy}^2]}\}/2\mu S_{xy}. \qquad (L8.10)$$

▶ *Exercise* L8.1.2 Show that the other root of (L8.9) can be rejected because it does not give the correct result when $\mu = 0$. □

Actually, formula (L8.10) for a_2 is numerically unstable because of subtractive cancellation, especially when the relative error in x to that in y, and thus μ, is small. I learned this the hard way by coding this form originally in the programs in L8.2, then noticing that the slope was very sensitive to μ when μ was small. However, Problem [A4.3] (*c*) shows how to remedy the subtractive cancellation that arises. By rationalizing the numerator you will readily find the stable formula for the slope, which is reliable for all μ,

$$a_2 = 2S_{xy}/\{\sqrt{[(S_{xx} - \mu S_{yy})^2 + 4\mu S_{xy}^2]} + (S_{xx} - \mu S_{yy})\}. \qquad (L8.11)$$

The minimum χ^2 value is readily computed formally by substituting into the definition (L8.1), with the simplification (L8.5), and noting that the sums have already been computed, to produce simply

$$\chi^2_{min} = S_{yy} - 2a_2 S_{xy} + a_2^2 S_{xx}. \qquad (L8.12)$$

Note that the value of χ^2 thus obtained depends on the choice of scale for the weight at each point.

▶ *Exercise* L8.1.3 Show all the steps in deriving (L8.12) for the minimum χ^2. □

With all this mathematical machinery set in motion, you should be eager to get some useful computer programs running to make least-squares fits.

L8.2 {1} Sample programs for straight-line fitting

The straight-line weighted least-squares fit program, LSFCHK, provided here will enable you to explore analyses of a variety of data, such as those presented in section L8.3. The programs allow both the x and the y variables to have uncertainties, provided that the uncertainties are in the same ratio for all the data points. The relation between the x and y errors and the variables μ is given in (L8.3). The y variables themselves are assumed to have weights w_k, which vary as k varies from 1 to the number of data points n. Usually these weights would be inversely proportional to the square of the error in y at each point. The formula for the slope a_2 is (L8.11), that for the intercept a_1 is (L8.6), and the minimum χ^2 is obtained from (L8.12).

The structure of the least-squares checking program LSFCHK and the procedure SLWLSF (Straight Line Weighted Least Squares Fit) is very straightforward, so I have omitted a pseudocode outline. The programs have been coded in both Pascal and Fortran. If you want a BASIC program, transliterate one of these programs.

Straight-line fitting in Pascal

The Pascal version of the least-squares fit program contains a single procedure, SLWLSF, and the driver program, LSFCHK, which defines the global variables, performs the input and output in a simple format, and invokes SLWLSF.

If you intend to use the programs extensively, then you should modify the input and output control in a way that is suitable for your computer system, in order to check for variables out of range and to allow more flexible use of the program. The code presented here runs correctly on an Apple II +, using only Pascal statements compatible with UCSD Pascal, so it should be suitable for a wide range of microcomputers and workstations.

```
PROGRAM LSFCHK;
(* Straight line least squares fit, including
   weights, X and Y errors; checking program *)
USES TRANSCEND; (* for SQRT *)
VAR K,N:INTEGER;
    MU,A1,A2,XAV,YAV,CHISQD:REAL;
    XLIST,YLIST,WGHT:ARRAY[1..100] OF REAL;
PROCEDURE SLWLSF;
(* Straight line weighted least squares fit,
   including X and Y errors *)
VAR SW,SX,SY,XK,YK,SXX,SXY,SYY,QUAD:REAL;
BEGIN
 (* Weighted averages of X and Y *)
 SW:=0; SX:=0; SY:=0;
 FOR K:=1 TO N DO
 BEGIN
  SW:=SW+WGHT[K]; (* summing weights *)
  SX:=SX+XLIST[K]*WGHT[K];
```

```
 SY:=SY+YLIST[K]*WGHT[K];
END;
XAV:=SX/SW; YAV:=SY/SW;
(* Weighted bilinear sums *)
SXX:=0; SXY:=0; SYY:=0;
FOR K:=1 TO N DO
BEGIN
 XK:=XLIST[K]-XAV; YK:=YLIST[K]-YAV;
 SXX:=SXX+XK*XK*WGHT[K];
 SXY:=SXY+XK*YK*WGHT[K];
 SYY:=SYY+YK*YK*WGHT[K];
END;
(* Slope, A2 *)
QUAD:=SXX-MU*SYY;
A2:=2*SXY/(SQRT(QUAD*QUAD+4*MU*SXY*SXY)+QUAD);
(* Intercept, A1 *)
A1:=YAV-A2*XAV;
(* Minimum chi-squared *)
CHISQD:=SYY-2*A2*SXY+A2*A2*SXX;
END; (* SLWLSF *)
BEGIN
 WRITELN('Straight Line Least Squares; Pascal');
 WRITE('Number of data, N?'); READLN(N);
 WRITELN('Input ',N:2,' triples of X,Y,W:');
 FOR K:=1 TO N DO
 READLN(XLIST[K],YLIST[K],WGHT[K]);
 WRITELN('Ratio of X to Y variances, MU?');
 READLN(MU);
 SLWLSF; (* Weighted fit to a straight line *)
 WRITELN('Intercept, A1=',A1:8,
         ': Slope,A2=',A2:8);
 WRITELN(' CHISQD=',CHISQD:8);
END.
```

This Pascal program was tested by feeding in nearly exact straight lines and verifying that the slope and intercept agreed with the externally calculated values. The option for x as well as y errors ($\mu \neq 0$) was checked by running small values of μ and comparing the results to other least-squares programs and also by running cases with input data having known slope and with $\mu = 1$, for which symmetry guarantees reciprocal slope for y on x and for x on y.

Fitting to straight lines from Fortran

I coded the Fortran version of the program LSFCHK, which contains the least-squares fitting subroutine SLWLSF (Straight Line Weighted Least

Squares Fit) by editing the Pascal version. This is tedious, but not error prone if one has a good editor. The Pascal global variables were made into arguments passed to the least-squares subroutine through the argument list in the Fortran CALL. I achieved a correct program after two compilations, a record for this combination of hardware, software, and wetware.

The Fortran straight-line fitting program was tested on the Apple II + microcomputer using Apple Fortran, and it ran (with changes needed only in the input/output unit references) in WATFIV on a mainframe computer. The results from the Fortran systems and the Pascal version agree.

```
      PROGRAM LSFCHK
C  Straight line least squares fit, including
C  weights, X and Y errors; checking program
      INTEGER N,K
      REAL MU,A1,A2,CHISQD,
     1     XLIST(100),YLIST(100),WGHT(100)
C  A * denotes terminal output or input
   10 WRITE(*,12)
   12 FORMAT(' Straight Line Least Squares Fit;',
     1' Fortran')
      WRITE(*,14)
   14 FORMAT(' Number of data, N?')
      READ(*,16) N
   16 FORMAT(I2)
      WRITE(*,18) N
   18 FORMAT(' Input',I2,' triples of X,Y,W:')
      READ(*,20)(XLIST(K),YLIST(K),WGHT(K),K=1,N)
   20 FORMAT(3F5.2)
      WRITE(*,22)
   22 FORMAT(' Ratio of X to Y variances, MU?')
      READ(*,24) MU
   24 FORMAT(F5.2)
C  Weighted fit to a straight line
      CALL SLWLSF(N,MU,XLIST,YLIST,WGHT,
     1            A1,A2,CHISQD)
      WRITE(*,26) A1,A2,CHISQD
   26 FORMAT(' Intercept, A1=',E10.3,
     1': Slope, A2=',E10.3,': CHISQD=',E10.3)
      GO TO 10
      END
      SUBROUTINE SLWLSF(N,MU,XLIST,YLIST,WGHT,
     1                  A1,A2,CHISQD)
C  Straight line least squares fit,
C  including weights, X and Y errors
      INTEGER N
      REAL SW,SX,SY,XK,YK,SXX,SXY,SYY,QUAD,
```

```
     1        MU,A1,A2,CHISQD,
     2        XLIST(100),YLIST(100),WGHT(100)
C  Weighted averages of X and Y
      SW = 0
      SX = 0
      SY = 0
      DO 100 K = 1,N
C  Summing weights
      SW = SW+WGHT(K)
C  also X and Y values
      SX = SX+XLIST(K)*WGHT(K)
      SY = SY+YLIST(K)*WGHT(K)
  100 CONTINUE
      XAV = SX/SW
      YAV = SY/SW
C  Weighted bilinear sums
      SXX = 0
      SXY = 0
      SYY = 0
      DO 200 K = 1,N
      XK = XLIST(K)-XAV
      YK = YLIST(K)-YAV
      SXX = SXX+XK*XK*WGHT(K)
      SXY = SXY+XK*YK*WGHT(K)
      SYY = SYY+YK*YK*WGHT(K)
  200 CONTINUE
C  Slope, A2
      QUAD = SXX-MU*SYY
      A2 = 2*SXY/(SQRT(QUAD*QUAD+4*MU*SXY*SXY)
     1        QUAD)
C  Intercept, A1
      A1 = YAV-A2*XAV
C  Minimum chi-squared
      CHISQD = SYY-2*A2*SXY+A2*A2*SXX
      RETURN
      END
```

The exercises in the next section give you a wide variety of data sets from research and development in science and engineering. Using these data you can test straight-line hypotheses on the relations between pairs of variables.

L8.3 {1} Quarks, radiocarbon dating, solar cells, and warfare

The title of this section indicates that least-squares fitting to straight lines has a broad appeal in science. The elementary-particle physicist uses it to test the hypothesis that there are particles ("quarks") whose electric charge is a fraction of that of the electron; this is part of the quark model of fundamental particles. The archaeologist is interested in least squares when checking the radiocarbon time scale against the ages of Egyptian antiquities. For the space scientist and electrical engineer the concern is in predicting the degradation of solar-cell performance in space. The political scientist and historian use least-squares fitting, referred to as "regression lines," in their studies of whether the human race is becoming more warlike.

Evidence for fractional charges

For the first half of the twentieth century there was firm experimental evidence that the smallest nonzero charge was that of the electron and that all other charges were integral multiples of the magnitude of the electron charge, e. From the middle 1960s onward, theoretical physicists postulated fractionally charged fundamental particles, called *quarks*, combinations of which would make up the constituents of nuclei, namely mesons, neutrons, and protons. Other particles familiar to high-energy physicists could also be described by the quark model. You can read about this research in the *Scientific American* article by Schwitters on fundamental particles with charm.

However, no isolated quark had been discovered, in spite of several intensive experimental efforts. In the mid-1970s refinements of experiments of the Millikan oil-droplet type, but done with tiny metal spheres, were made in order to search for free fractional charges. Data from a series of experiments made at Stanford University, as reported in the *Physical Review Letters* research journal in 1981 by LaRue and coworkers, are summarized here. Their results were expressed as

$$q/e = q_r + n, \tag{L8.13}$$

where q is the charge on the metal sphere, e is the electron charge, q_r is called the *residual charge* (measured in units of e) and n is integral. The data are presented by the order of the spheres used. In some cases repeat measurements of the spheres were made after altering conditions, for example, by heating them.

If there are no fractional charges, then the residual charge q_r should be zero. Thus, a least-squares fit of q_r against measurement number k, should be a straight line with zero intercept and zero slope. Table L8.3.1 shows the data from the 1981 report by the Stanford researchers.

With these data and the least-squares program LSFCHK you should be ready to investigate the evidence for fractional charges.

▶ *Exercise* L8.3.1   (Least-squares analysis of fractional charge data)
(a) Make a plot of the residual charge, q_r, against measurement number, k. Indicate the errors in q_r by bars above and below each data point. For the plotting it is

Table L8.3.1   Residual charges on metal spheres.

k	q_r	e_k	k	q_r	e_k	k	q_r	e_k
1	0.000	0.03	14	− 0.030	0.03	27	+ 0.365	0.03
2	+ 0.090	0.08	15	− 0.030	0.03	28	0.000	0.01
3	− 0.335	0.08	16	+ 0.030	0.03	29	+ 0.365	0.03
4	− 0.015	0.03	17	+ 0.335	0.04	30	0.000	0.01
5	− 0.015	0.06	18	+ 0.380	0.04	31	+ 0.015	0.02
6	+ 0.060	0.09	19	− 0.380	0.05	32	− 0.030	0.03
7	− 0.030	0.09	20	+ 0.015	0.02	33	0.000	0.01
8	+ 0.320	0.02	21	+ 0.275	0.02	34	− 0.320	0.01
9	+ 0.030	0.03	22	+ 0.015	0.02	35	0.000	0.01
10	0.000	0.03	23	− 0.365	0.02	36	0.000	0.02
11	+ 0.335	0.02	24	− 0.015	0.01	37	− 0.380	0.03
12	− 0.015	0.03	25	+ 0.040	0.02	38	+ 0.380	0.06
13	+ 0.300	0.03	26	+ 0.010	0.01	39	+ 0.015	0.02

convenient to use the simple graphics programs given in L4. Are there any anomalous points? If so, what do they imply?

(b) To obtain an objective analysis of the data make a straight-line weighted least-squares fit of q_r against k, using the program LSFCHK from section L8.2. For weights use the reciprocal squares of the errors from Table L8.3.1. The uncertainty in the measurement number is zero, so μ = 0 in LSFCHK. The input and output formats in LSFCHK should be modified to accommodate the number of significant figures in the data. What is unrealistic about this analysis?

(c) Omit measurements that have residual charges differing by much more than their errors from the rest of the data. Repeat the analysis of part (b). What do you interpret for the intercept and slope of the least-squares fit?

(d) The meaningfulness of fitting a functional relationship between q_r and k is questionable, because the time sequence (indicated by measurement number) is probably not relevant, unless the experimental design is faulty. Modify LSFCHK to calculate weighted standard deviations among the values of q_r, both with and without the anomalous measurements. The program should output the averages (which are already calculated) and the standard deviations. Our data sample differs very slightly from that used by LaRue; do you get 0.001 ± 0.003 for the residual charges that are near zero, and − 0.343 ± 0.011 and + 0.328 ± 0.007 for the charges that have magnitudes near 0.3? What simple fraction of an electron charge would you guess for the "quark"? □

It will be interesting for you to follow up on these and other experiments relating to quarks. Have the Stanford experiments and their interpretation withstood the scientific challenges of other experiments and other interpretations?

Radiocarbon dating and Egyptian antiquities

The nuclei of ^{14}C are unstable, decaying with a half-life of 5730 years. Plant life incorporates ^{14}C from the atmosphere during growth, and its radioactivity decays exponentially after the death of the plant. From the current decay rate of a sample of the plant material we can therefore deduce the time when the plant was growing. This radiocarbon-dating technique is sensitive for objects with an age comparable to the half-life.

However, the technique of radiocarbon dating has its problems. For example, one must know the atmospheric concentration of ^{14}C at the time the plant was growing. One might assume that it was the same as the current concentration. One way of checking the radiocarbon dating time scale is to use the technique to date objects of known antiquity, then to compare the two results. A discrepancy may be attributed either to the antiquarian or archaeologist responsible for the historical date, or to the radiocarbon technique, to both, or to forgery of the antique.

In his book *Physics and Archaeology* M. J. Aitken presents data on radiocarbon results for historically dated samples of Eqyptian antiquities, mostly obtained from royal tombs of reliably established age. The selection of data that we show in Table L8.3.2 is from his Table 2.1. The data are for reeds, which would likely be used soon after they were cut. Aitken's table also shows data for wood, which, in unforested Egypt, would likely have been used several times after its parent tree had been cut down and before it was used in building a tomb. Therefore, dating from wood samples is probably less reliable than from reeds.

Table L8.3.2   Radiocarbon dates for Egyptian reed samples

h	r	h	r
− 3100	− 2580	− 1890	− 1770
− 2950	− 2485	− 1255	− 1180
− 2650	− 2215	− 650	− 610
− 2350	− 2005	− 370	− 445

In the table the historical date, h, is in years in the Western calendar, as is the radiocarbon date, r. The uncertainty in each of the historical dates is about 20 years, and the uncertainty in each of the radiocarbon dates is about 45 years. Therefore, a realistic estimate of the relative uncertainty in the historical date to the radiocarbon date is $\mu = (20/45)^2 \approx 0.20$. Each of the data values is equally significant, therefore the same weight factors (say 1 for convenience) should be used for each point.

▶ *Exercise* L8.3.2   (Regression analysis of radiocarbon and historical dates)
(a) Use the program LSFCHK to fit the relation between h and r using the data from

Table L8.3.2, assuming that the relationship is linear. Let the historical date be the independent variable and the radiocarbon date the dependent variable. The input and output formats of LSFCHK should be modified for ease of input and readability of output.

(b) Check whether the intercept and slope in your analysis are sensitive to the value assumed for the relative errors in h and r by running LSFCHK with both $\mu = 0$ (relatively negligible error in h) and with $\mu = 0.20$, the value indicated as appropriate for these data. Then, doubting the Egyptologists, try $\mu = 1.0$, which corresponds to a random error of about 45 years in each of the antiquity dates. (An error of 45 years corresponds to about life expectancy in ancient Egypt.) Do you think that the differences in slope are significant?

(c) Reanalyze the data by fitting h on r, rather than r on h. Note that you must also invert the value of μ. Does the slope come out the reciprocal of that in part (a)? Are any differences significant, in view of the uncertainties in the data?

(d) Modify the least-squares program by adding a program segment that predicts h given r and the intercept and slope from part (c). Use it to calculate the standard deviation between the radiocarbon dates given in Table L8.3.2 and those predicted by the regression line, using the historical dates from the same table.

(e) Make a graph of the difference $h - r$ against r, for r in the range 0 to 5000 B.C. Note that use of $h - r$ rather than r makes the display less sensitive to the resolution of the display device. For example, you might use the simple printer plotter from L4, since its resolution would be well within the 50 year uncertainty in any determinations of age.

(f) Suggest explanations of the source of the discrepancies between the historical and radiocarbon dates. As a help, consult Aitken's book on physics and archaelogy. □

After this rather dated excursion into the gloomy catacombs of ancient history, you would probably be glad to make a sunny excursion into outer space.

Solar cell efficiency in space

You will be aware of the developing technology of photovoltaic energy conversion from sunlight. Much of the research in this area has been done as part of the space research program, because energy sources other than sunlight are expensive to transport to a space vehicle and are of limited lifetime.

For applications in space, photovoltaic cells must have excellent long-term performance, both because of the cost of lifting the system into orbit and because of the more difficult maintenance conditions in space compared with those on Earth. Development of solar photovoltaic power sources for space applications should also result in lower costs and improved engineering performance for Earth-based systems, such as for electrical utility systems, both large-scale power plants and single-house systems.

In his *Solar Cell Array Design Handbook*, H. S. Rauschenbach presents extensive data on photovoltaic energy conversion devices. In particular, he has investigated the performance degradation of solar cells in the near-space

Table L8.3.3   Photovoltaic cell power degradation

D	P	D	P
0	67	115	60
45	64	170	60
80	62	210	58

environment. We will use the data from his Figure 4-55 for the OCLI violet solar cell power output per unit area, P, as a function of days in orbit, D, in order to predict the useful lifetime of such a photovoltaic array. The perform-ance data are given in Table L8.3.3, in which the power is in mW/4cm².

▶ *Exercise L8.3.3*   (Degrading performance of solar cells)
(*a*) Make a straight-line least-squares analysis of the data in Table L8.3.3, with D as the independent variable and P as the dependent variable. Assume equal weight for each of the power measurements, and negligible error in D.
(*b*) What is the initial power output of the OCLI cell, according to the least-squares analysis? After how many days in orbit will this photocell produce only about half as much power as originally?
(*c*) Suppose that a communications satellite equipped with OCLI solar cells has been in orbit 3 years. To run the cell circuitry requires 10% of the initial power output, and it is not considered feasible to operate the solar panels if the useable power out-put is less than 10% of the original output. It takes 50 days to arrange for refurbish-ment by a space shuttle. How long from now should the space-shuttle company be contacted? □

Now that you have considered the sunny side for sources of electrical power, ponder a little on the next exercise on sociology and political power.

Is war on the increase?

Regrettably, war involves applied science. There is a general impression that war is on the increase throughout the world. But, by some realistic measure, is this so? The exercises to follow will enable you to investigate this question.
    In *The Wages of War* 1816–1965: *A Statistical Handbook*, David Singer and Mel-vin Small provide much data relating to secular trends in the incidence of war. One measure of bellicosity is the number of battle deaths per million population, D, in each war. Data on this are given in Table 4.2, page 66, in Singer and Small. Our Table 8.3.4 is a subset of their table, selected for inter-state wars at roughly 5-year intervals. The year, Y, given for each war is about the middle year of the war.
    From these grim statistics we can investigate whether there is firm evidence that war is on the rise.

Table L8.3.4   Battle deaths from interstate wars 1823–1956.

Y	D	Y	D	Y	D
1823	23	1866	384	1916	14137
1829	1660	1877	2336	1921	3650
1848	230	1885	26	1934	38235
1849	29	1895	31	1939	1648
1854	1586	1898	106	1942	10912
1859	292	1906	270	1948	260
1860	40	1910	408	1956	153

► *Exercise L8.3.4*   (Least-squares analysis of battle deaths)

(a) Make a regression line analysis (straight-line least-squares fit) of lg (D) from Table L8.3.4 against the year Y of wars of the past two centuries. Assume that all wars have equal weight and that all the data are equally reliable. From the slope of the fit obtained using program LSFCHK from L8.2, is there a significant rate of increase in war as measured by the quantity D? If you obtain a positive slope, when would your least-squares fit retrodict that there should have been essentially no wars? Does this agree with the historical record? If you obtain a negative slope, when do you predict that wars will practically be at an end?

(b) Omit at random some of the wars listed in Table L8.3.4. To do this, generate a table of random numbers between 1 and 21, then use its first half dozen entries to decide which numbers in the table to eliminate. (You could use the methods in computing laboratory L6.1 to make the random numbers.) Does the slope of the regression line change when fewer entries are used? As a check, compare the values of χ^2/N for $N = 21$ (the number of data points in Table A8.3.4) and for $N = 15$ (the number of data points reduced by 6).

(c) Don't most people think that war is becoming more common? Give reasons why this should be so. As a scientist, what can you do to reduce warfare? (For background reading, specifically on the nuclear arms race, see the resource article by Schroeer and Dowling.) □

You can further your understanding of least-squares analysis of data from the readings following. In particular, the book by Daniel, Wood, and Gorman gives many practical programmed examples from applied science, including both nonlinear and linear least squares analyses. The book edited by Davies and Goldsmith has many examples from industrial chemistry and that written by Hovanessian introduces the engineering-oriented reader to techniques of filtering. General techniques for exploring data and their representations are suggested in Tukey's book.

A minimal reading list on least squares

Aitken, M. J., *Physics and Archaeology*, Clarendon Press, Oxford, 1974.
Daniel, C., F. S. Wood, and J. W. Gorman, *Fitting Equations to Data*, Wiley, New York, 1980.

Davies, O. L., and P. L. Goldsmith (eds.), *Statistical Methods in Research and Production*, Imperial Chemical Industries, 1976.

Hovanessian, S. A., *Computational Mathematics in Engineering*, D. C. Heath, Lexington, Mass., 1976.

LaRue, G. S., J. D. Phillips, and W. M. Fairbank, "Observation of Fractional Charge of 1/3)e on Matter," *Physical Review Letters*, **46**, 967 (1981).

Pasachoff, J. M., "Applicability of Least-Squares Formula," *American Journal of Physics*, **48**, 800 (1980).

Rauschenbach, H. S., *Solar Cell Array Design Handbook*, Van Nostrand, New York, 1980.

Schroeer, D., and J. Dowling, "Resource Letter PNAR-1: Physics and the Nuclear Arms Race," *American Journal of Physics*, **50**, 786 (1982).

Schwitters, R. F, "Fundamental Particles with Charm," *Scientific American*, October 1977, p.56.

Singer, J. D., and M. Small, *The Wages of War*, 1816–1965, Wiley, New York, 1972.

Tukey, J. W., *Exploratory Data Analysis*, Addison Wesley, Reading, Mass., 1977.

L9 FOURIER ANALYSIS OF AN EEG

This computing laboratory is an introduction to practical Fourier analyses. It shows how the discrete Fourier transform theory from A5.2 is applied to the analysis of real data. The data suggested for analysis are from human brain waves. Such a brain-wave pattern is technically known as an *electroencephalogram*, or *EEG*. This laboratory also gives you an opportunity to develop and use the fast Fourier transform (FFT) algorithm derived in section A5.8. The FFT may then be used to compute the discrete Fourier transform of the EEG data. Other computerized applications of Fourier expansions are given in Merrill's book of programming exercises in physics, in the article by Schmidt on Fourier analysis using a pocket calculator, and in the *Byte* article by Stanley and Peterson about FFT on small computers.

The Fourier analysis of an EEG is an example of an analysis of data in which, to the untrained observer, the voltage output as a function of time has very little pattern. The analysis of the EEG data transforms it from the time domain to the frequency domain, in which the dominance of a few frequencies becomes evident, as you will discover in L9.2. We showed in A4.7 and A5.1 that Fourier amplitudes provide the best fit to the data, in the least-squares sense, that can be obtained using cosine and sine functions or complex exponentials.

Data from many other physiological rhythms may also be analyzed by the Fourier analysis methods developed in this computing laboratory. Among several examples discussed in Cameron and Skofronick's book on medical physics are magnetoencephalograms and magnetocardiograms, which require measurement of magnetic fields of about 10^{-9} Gauss by the use of ultrasensitive magnetometers. (See Cameron and Skofronick, Chapter 9.7.)

The Fourier analysis methods that we use here are useful in the frequency analysis of any complicated waveform. For example, the seismograph depicts the mechanical vibrations of the Earth and other heavenly bodies and portrays their inner rumblings. The analysis of such geophysical data is detailed in the textbook by Robinson and Treitel and in the *Scientific American* article by Boore. You can read about the impact of computer technology on the monitoring of the volcanic activity of Mount St. Helens in the U.S. Northwest in the article, "Cataclysmic Computing," by J. Rose in *Datamation* magazine. The analysis of time series by Fourier expansion methods is described in the book by Otnes and Enochson, which also has many Fortran programs for such analyses.

This laboratory begins with an overview of electroencephalography in L9.1, then in L9.2 the calculation of the frequency spectrum is outlined, together with its use to regenerate the EEG voltage. In section L9.3 we are concerned with the Nyquist sampling criterion and the effects of noise. The autocorrelation, which provides an alternative characterization of the data, is introduced in L9.4, where methods for its calculation are described. Extensive references to practical Fourier analyses and their application to EEG and other data are presented in the concluding section.

L9.1 {2}   Introduction to encephalography

In this section we provide an overview of the EEG. Then we present the clinical record for a sample human EEG and outline the program for its Fourier analysis and for other analysis techniques.

What is an EEG?

The variation of potential differences between points on the scalp as a function of time have, since the 1920s, been associated with the electrical activity of the brain. The early measurements, summarized in the articles by Walter and by Brazier, were subject to much controversy. In part this was because the direct interpretation of the potentials, V, as a function of time, t, was very subjective. The advent of computers in the 1950s enabled objective analysis of brain waves, especially their frequency components. An electroencephalogram (EEG) is just the trace of the voltage pattern $V(t)$.

   For an adult human a typical EEG has the form shown in Figure L9.1.1. The main frequencies in an adult human EEG are 0.3 to 3.5 Hz (called δ waves, predominating in sleeping adults and in awake young children), 8 to 13 Hz (called a waves, predominant in awake, relaxed adults), and 18 to 30 Hz (called β waves, appearing sporadically during sleep). The interpretation of such patterns, especially for diagnosis of neural dysfunction, is discussed at an introductory level by Cameron and Skofronick, and by Hobbie. Instrumentation for EEG data acquisition is described in the book by Cromwell, Weibell, and Pfeiffer. Techniques of EEG analysis are presented in the book by Ledley, while the monograph by Spehlmann provides many

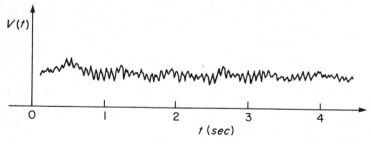

Figure L9.1.1   EEG pattern of a healthy, relaxed science student. The predominance of a waves at about 10 Hz is evident.

examples of EEG traces and their clinical interpretation. The study of brain function by the analysis of EEGs from evoked responses to external stimuli is described in a *Scientific American* article by Regan.

The relation between computers and brains is the theme of novels by Crichton and by Ryan. One of the pioneers of the digital computer, J. Von Neumann, wrote a thoughtful book on the relation between the computer and the brain.

Salmon EEG and socioeconomics

The EEG patterns of animals other than humans are also studied extensively. For example, while studying the physiological and behavioral mechanisms in migratory fish, especially the homing mechanisms of salmon, Hasler and coworkers reported large increases in EEG amplitudes when trace odors of the chemical morpholine were detected by salmon that had been imprinted with it when young. From this research there have been developed chemical imprinting techniques for persuading hatchery fish to return to their original hatchery so that they can be easily caught after their ocean migration. This salmon ranching is of great practical value in the salmon hatchery and fishing industry. Hasler's original basic research has thus reaped unexpected practical rewards.

A sociological and economic consequence of the success of large-scale salmon ranching is that traditional, small-scale salmon fishermen, especially those in the U.S. Northwest, are unlikely to be able to compete economically with the large companies that build and run the hatcheries. Thus, basic research may have unexpected social costs.

The clinical record

Many EEG patterns of a patient, for confidentiality referred to as NBI (No-one But Initials), have been recorded. We consider, for illustration, only a very short section of these records, namely a 1-sec interval. For our purposes we use simulated EEG patterns for which the amplitudes of the various frequencies are characteristic of those of human adults, as discussed earlier. Noise in unknown amounts is also present in the EEG records.

The EEG data are in digital form, with 32 data points running from 0 to 1 sec in 1/31 sec intervals. To achieve a periodic function, the first and last points are the same. Table L9.1.1 provides three representative data sets, each differing in the amount of noise present:

▶ *Exercise* L9.1.1   Choose one of the three sets of EEG data in Table L9.1 and display it similarly to Figure L9.1.1. Estimate the predominant frequency components by inspection of this raw record. Explain and justify your estimation method. □

Table L9.1.1   EEG data sets; Time = (IT − 1)/31 sec; voltages in μV.

IT	$V_1(t)$	$V_2(t)$	$V_3(t)$
1	13.93	12.07	13.00
2	20.77	19.56	19.64
3	− 21.62	− 12.14	− 17.18
4	− 19.28	− 11.45	− 14.66
5	28.15	29.75	29.42
6	23.04	22.18	21.61
7	− 32.87	− 32.32	− 33.07
8	− 29.30	− 38.27	− 32.49
9	17.61	18.82	18.51
10	18.63	26.64	21.16
11	− 14.58	− 17.42	− 16.00
12	0.38	− 8.74	− 2.71
13	45.21	35.97	40.30
14	20.91	27.56	22.94
15	− 30.27	− 27.19	− 28.26
16	− 25.33	− 26.71	− 25.02
17	5.17	9.14	6.68
18	− 3.66	− 7.36	− 6.22
19	− 28.17	− 26.13	− 26.86
20	0.97	6.47	4.25
21	35.69	30.31	33.00
22	3.56	1.79	2.15
23	− 28.18	− 30.39	− 29.58
24	− 11.30	− 8.54	− 9.21
25	14.25	10.63	12.94
26	4.56	− 1.78	0.39
27	− 0.48	− 6.38	− 3.90
28	14.99	16.67	17.13
29	19.68	22.84	21.56
30	− 2.17	1.69	− 1.71
31	− 16.35	− 20.65	− 18.00
32	13.93	12.07	13.00

From the analysis of EEG records it is believed that the largest Fourier amplitudes are obtained for frequencies k near 3 and 10 Hz, with frequencies above 12 Hz having relatively small amplitudes. As in all bioexperimentation, there is probably a large amount of noise present in the EEG records coming from the patient and the clinical environment. In section L9.3 of our analysis we will investigate the magnitude of this noise.

Fourier analysis program structure

Here is a suggested structure for your program to Fourier analyze an EEG. We outline the program structure in pseudocode:

L9 Fourier analysis of an EEG

> Data and options input
> L9.2 Frequency spectrum analysis of the EEG
> > by discrete Fourier transform:
> > Loop over frequencies k
> > > For each k, sum over all times t
> >
> > Prediction of voltages from Fourier ampltudes:
> > > For each time t, sum over all frequencies k
>
> L9.3 The Nyquist criterion and noise:
> > Compute standard deviation between input
> > > and predicted voltages
> > Double data point spacing and reanalyze (Option)
> > Add noise to data and reanalyze (Option)
>
> L9.4 Autocorrelation analysis of the EEG: (Option)
> > Loop over time lag TAU
> > > For each TAU integrate over all times t
>
> Go for more data or options
>
> Program end.

From this outline you can build up the detail of the program, including the use of input options to select which segments of the program are to be executed. Some details of implementing the program are given in the following sections.

L9.2 {2}   Frequency spectrum analysis of the EEG

We want (or are required) to first compute the Fourier amplitudes a_k and b_k in the discrete Fourier transform discussed in A5.2. From that section we have that

$$V(t) \; = \; \Sigma \, [a_k \cos (2\pi k t) \; + \; b_k \sin (2\pi k t)], \qquad (L9.1)$$

in which the sum runs over frequency k from 1 to 32 Hz. Correspondingly, t

has 32 values from 0 to 1 sec by 1/31 sec intervals. Note that we are making a discrete Fourier transform. Its distinction from other Fourier expansions is clarified in A5.1.

The formulas for the Fourier coefficients are derived in A5.2. To compute these coefficients we suggest using initially the conventional Fourier transform (CFT), discussed in A5.2 and A5.8, rather than the more efficient fast Fourier transform (FFT), introduced in A5.8. However, if your EEG Fourier analysis program is carefully designed, you will be able to use the two algorithms interchangeably.

In making EEG traces, the recorder is sometimes set so that only the fluctuations in the voltage are recorded and not any DC level.

(Note that in Figure L9.1.1 the voltage origin is not specified.) Therefore the coefficient a_0 in each of the data sets in Table L9.1.1 is assumed to be zero, and the average value of the EEG voltage is correspondingly zero.

▶ *Exercise* L9.2.1   (Averages and Fourier amplitudes)
(*a*) Show that the direct-current (DC) voltage level of a time-dependent signal, such as an EEG voltage, is described by a_0.
(*b*) Derive analytically the result that the average of a voltage over any time period that is a multiple of the fundamental period is given by its Fourier-expansion amplitude a_0.
(*c*) Check how closely the data that you have selected from Table L9.1.1 satisfy the assumption that their time average is zero. How can you modify data, if necessary, to achieve this? ☐

Our EEG analysis therefore requires the Fourier coefficients only for $k > 0$, which are given by equations (A5.16) and (A5.18). The coefficients of the cosine terms are

$$a_k = \Sigma \, V_j \cos (2\pi kj/N)/\sqrt{N}, \tag{L9.2}$$

and the coefficients of the sine terms are

$$b_k = \Sigma \, V_j \sin (2\pi kj/N)/\sqrt{N}. \tag{L9.3}$$

In these two equations the sums are over j from 1 to N (the number of data points), and V_j is the voltage at the jth time interval. The value of N should be input to your program as a variable, say NDATA, because it will be varied in examining the Nyquist criterion and the effects of noise, as described in section L9.2.

Coding the Fourier amplitude calculation

In this subsection you will find clues for coding, displaying, and interpreting the Fourier amplitudes (frequency spectrum) of an EEG. The pseudocode

outline at the end of L9.1 should serve as your guide in writing a well-struc-
tured program.

You will find it most convenient in the long run to write your program in
modular form and to provide input options to select the appropriate modules
for execution. This will help you in checking out the program, and it will help
you and other users of the program to understand it once it is executing satisfactorily.

▶ *Exercise L9.2.2*   (Fourier transform spectrum)

Write a program segment that inputs an EEG data sample, as given in
Table L9.1.1, then computes its discrete Fourier transform coefficients a_k and b_k
according to the formulas (L9.2) and (L9.3). The EEG data from Table L9.1.1 and
the resulting amplitudes should be stored in arrays. The output from this program
segment should be a table in which the columns are labeled by k (in Hz), by a_k (in
μV) and by b_k (in μV).

There are two possibilities for the algorithm to compute the Fourier coefficients:
(*a*) Use the conventional Fourier transform (CFT) method, in which equations
(L9.2) and (L9.3) are coded directly as a pair of nested DO loops. The outer loop
will be over the frequency variable k, and its inner loop will be over the time index j.
(*b*) As an alternative to the CFT analysis, use the FFT programs in the next subsection if you are programming in Pascal or Fortran, or a library procedure for FFT if
your computer system has it, or a BASIC program such as that listed in Stanley and
Peterson's article. There is also a Pascal program for the FFT given in Cooper's
book. Thus you could use the FFT algorithm for the discrete Fourier transform of
your EEG data to compute the a_k and b_k. Note that our FFT procedure assumes that
the input data are complex numbers, and it produces complex Fourier amplitudes
c_k. This is no real problem, but you do have to be imaginative and set the imaginary
part of the input values to zero. Cooper discusses efficient methods for applying the
FFT to real data.

For the FFT option the number of data points should be a power of 2. Thus the
choice of 32 data points is very convenient because you can also analyze subsets of
size 16 and 8 (as well as 4 and 2, if you insist) with correspondingly coarser time
resolution. This will be useful in L9.3 when investigating the Nyquist criterion. In
this option your program should also output a labeled table of frequency and Fourier coefficients.
(*c*) Display the Fourier coefficients a_k and b_k against k, then interpret the results.
This can be done manually on a sheet of graph paper. Alternatively, if you have
completed L4, use the display program from there. Interpret the Fourier coefficients
analogously to the discussion in A5.3.

Add some coding to your program to compute the quantity $a_k^2 + b_k^2$, which is
proportional to the power or intensity contributed by frequency k. Display this
quantity plotted against k on the same graph as the amplitudes, but using different
plotting symbols. Interpret these results.
(*d*) Is the pattern of Fourier amplitudes as a function of k typical of that for a normal, adult human, as discussed in L9.1? Estimate what values of k are the ampli-

tudes meaningful, in the sense that they are probably insensitive to the time resolution of the data and to the presence of noise in the EEG signal? ☐

As a help for those who want to apply the fast Fourier transform (FFT) algorithm for calculating the discrete Fourier transform of the EEG data, the next subsection provides suitable program segments.

FFT programs in Pascal and Fortran

In section A5.8 we derived a fast Fourier transform (FFT) algorithm for computing discrete Fourier transforms and we showed by examples that it can be many orders of magnitude faster than the conventional Fourier transform (CFT) in practical application. Here we provide a procedure, FFT, and a driving program, FFTEST, that shows how FFT is used. The program consistency can be checked by applying the FFT to real data (array FR) that you input, then computing the inverse transformation on the result. As expected, the original data are reproduced to within computer accuracy. It is assumed in FFTEST that the number of data points, NF, is a power of 2, $NF = 2^\nu$, where ν is integral.

The program structure follows the outline given in A5.8 by computing discrete Fourier transforms of subdivided intervals, then reordering these by a bit-reversing procedure, BITREV, which does not depend on the particular computer system used. The algorithm for BITREV is that developed in Exercise A5.8.3.

Only real computer arithmetic, rather than the complex arithmetic in which the algorithm is described, is used in procedure FFT. For simplicity, the programs do not use tables of cosine and sine functions, which would make them execute faster. You could probably also make a faster program by using a computer-dependent bit-reversing routine coded in machine language, by using complex arithmetic, and by using table look-up for the circular functions.

Details of programming the FFT are given in Chapter 10 of Brigham's monograph on the FFT and in Cooper's book on Pascal for scientists. Our programs are given in both Pascal and Fortran coding. However, the two programs are made as similar as possible, which will enable you to compare the two languages. For BASIC versions of the FFT algorithm, consult the article by Stanley and Peterson, which also has several interesting examples.

The Pascal version of the FFT program is described first. It runs correctly on an Apple II + microcomputer, and it uses only Pascal statements compatible with UCSD Pascal. The source code of the functions COS, SIN, and SQRT is in the source library TRANSCEND. The program is self-testing, as described at the beginning of this subsection.

```
PROGRAM FFTEST;
(* Test Fast Fourier Transform self-consistency *)
USES TRANSCEND; (* for COS,SIN,SQRT *)
```

```
VAR NU,NF,I,K,J,IT:INTEGER;
   FR,FI:ARRAY[1..32] OF REAL;
FUNCTION BITREV(J,NU:INTEGER):INTEGER;
(* J has NU bits; reverse into BITREV *)
VAR JT,I,JD,BITEMP:INTEGER;
(* BITEMP avoids re-accessing BITREV *)
BEGIN
 JT:=J; BITEMP:=0;
 FOR I:=1 TO NU DO
 BEGIN
  JD:=JT DIV 2; BITEMP:=BITEMP*2+JT-2*JD; JT:=JD;
 END;
 BITREV:=BITEMP;
END; (* BITREV *)
PROCEDURE FFT;
(* F F T of ( FR([J]+iFI[J], J=1,NF ) *)
(* IT:= 1 for coefficients of exp(2PIikj/NF)
   IT:=-1 for inverse transformation *)
CONST TPI=6.283185;
(* Global variables are NU,NF,IT (INTEGER),
      FR,FI (ARRAYs of 1...NF of REAL) *)
VAR II,KNU,NU1,NSUB,I,J,NF1,ID,
   KNUP1,KN1,K,KSORT:INTEGER;
   TPN,ANGLE,WR,WI,TEMPR,TEMPI,NORM:REAL;
BEGIN
 TPN:=TPI/NF; II:=-IT;
 (* Set up for first subinterval *)
 KNU:=0; NU1:=NU-1; NSUB:=NF DIV 2;
 (* Loop over levels of length NF/2, NF/4,..1 *)
 FOR I:=1 TO NU DO
 BEGIN
  WHILE KNU < NF DO
  BEGIN
  (* Fourier transform subinterval *)
  FOR J:=1 TO NSUB DO
  BEGIN
   NF1:=KNU; FOR ID:=1 TO NU1 DO NF1:=NF1 DIV 2;
   ANGLE:=TPN*BITREV(NF1,NU);
   WR:=COS(ANGLE); WI:=II*SIN(ANGLE);
   KNUP1:=KNU+1; KN1:=KNUP1+NSUB;
   TEMPR:=FR[KN1]*WR-FI[KN1]*WI;
   TEMPI:=FR[KN1]*WI+FI[KN1]*WR;
   FR[KN1]:=FR[KNUP1]-TEMPR;
   FI[KN1]:=FI[KNUP1]-TEMPI;
   FR[KNUP1]:=FR[KNUP1]+TEMPR;
   FI[KNUP1]:=FI[KNUP1]+TEMPI;
   KNU:=KNU+1;
```

```
    END;
    KNU:=KNU+NSUB;
  END; (* WHILE loop *)
 (* Set up for next subinterval *)
 KNU:=0; NU1:=NU1-1; NSUB:=NSUB DIV 2;
 END;
 (* Reverse bit pattern and sort transform *)
 FOR K:=1 TO NF DO
 BEGIN
  KSORT:=BITREV(K-1,NU)+1;
  IF KSORT > K THEN
  BEGIN (* Swap *)
   TEMPR:=FR[K];FR[K]:=FR[KSORT];FR[KSORT]:=TEMPR;
   TEMPI:=FI[K]FI[K]:=FI[KSORT]; FI[KSORT]:=TEMPI;
  END;
 END;
 (* Normalize *)
 NORM:=1.0/SQRT(NF);
 FOR K:=1 TO NF DO
 BEGIN
  FR[K]:=NORM*FR[K]; FI[K]:=NORM*FI[K];
 END;
END; (* FFT *)
BEGIN
 WRITELN('FFT test; Pascal: NU then FR array');
 READLN(NU); NF:=1; FOR I:=1 TO NU DO NF:=2*NF;
 FOR J:=1 TO NF DO READLN(FR[J]);
 FOR J:=1 TO NF DO FI[J]:=0;
 IT:=1; FFT; (* F F T into FR+iFI *)
 FOR J:=1 TO NF DO WRITELN(FR[J],FI[J]);
 IT:=-1; FFT; (* Inverse into same arrays *)
 FOR J:=1 TO NF DO WRITELN(FR[J],FI[J]);
END.
```

The Fortran version of the fast Fourier transform algorithm is similar to this Pascal version. The Fortran program FFTEST was tested on an Apple II + microcomputer using Apple Fortran. The FFT program was also tested on a macrocomputer, using the WATFIV dialect of Fortran 77. The FFT program is self-testing, as described at the beginning of this subsection.

```
      PROGRAM FFTEST
C  Test Fast Fourier Transform self-consistency.
      INTEGER NU,J,NF
      REAL FR(32),FI(32)
C  A * denotes terminal output or input
      WRITE(*,102)
```

```
  102 FORMAT(/' FFT test; Fortran:',
     1'NU, (FR(J),J=1,2**NU) in (I1/16F5.2)')
      READ(*,104) NU
  104 FORMAT(I1)
      NF = 2**NU
      READ(*,106) (FR(J),J=1,NF)
  106 FORMAT(16F5.2)
      DO 120 J = 1,NF
         FI(J) = 0.0
  120 CONTINUE
C  Compute FFT back into FR + iFI
      CALL FFT(FR,FI,NU,1)
      WRITE(*,106) (FR(J),FI(J), J=1,NF)
C  Compute inverse FFT into FR + iFI
      CALL FFT(FR,FI,NU,-1)
      WRITE(*,106) (FR(J),FI(J), J=1,NF)
      STOP
      END
      SUBROUTINE FFT(FR,FI,NU,IT)
C  F F T of ( FR(J) + iFI(J), J = 1, 2**NU)
C  IT =  1 for coefficients of exp(2PIikj/NF)
C  IT = -1 for inverse transformation
      INTEGER NU,IT,NF,NSUB,NU1,KNU,I,J,
     1        KNUP1,KN1,K,KSORT,BITREV
      REAL FR(32),FI(32),WR,WI,TEMPR,TEMPI,II,
     1 TPN,TPI,ANGLE,NORM
      DATA TPI/6.283185/
      NF = 2**NU
      TPN = TPI/NF
      II = -IT
C  Set up for first subinterval
      KNU = 0
      NU1 = NU-1
      NSUB = NF/2
C  Loop over levels of length NF/2, (NF/2)/2,..1
      DO 200 I = 1, NU
   40    CONTINUE
C  Fourier transform the subinterval
         DO 100 J = 1, NSUB
            ANGLE =TPN*BITREV(KNU/2**NU1,NU)
            WR = COS(ANGLE)
            WI = II*SIN(ANGLE)
            KNUP1 = KNU+1
            KN1 = KNUP1+NSUB
            TEMPR = FR(KN1)*WR - FI(KN1)*WI
            TEMPI = FR(KN1)*WI + FI(KN1)*WR
```

```
            FR(KN1) = FR(KNUP1)-TEMPR
            FI(KN1) = FI(KNUP1)-TEMPI
            FR(KNUP1) = FR(KNUP1)+TEMPR
            FI(KNUP1) = FI(KNUP1)+TEMPI
            KNU = KNU+1
  100    CONTINUE
        KNU = KNU+NSUB
        IF (KNU.LT.NF) GOTO 40
C  Set up next subinterval
        KNU = 0
        NU1 = NU1-1
        NSUB = NSUB/2
  200 CONTINUE
C  Reverse bit pattern & sort transform
        DO 300 K = 1, NF
            KSORT = BITREV(K-1, NU) + 1
            IF (KSORT.GT.K) THEN
C  Swap
                TEMPR = FR(K)
                FR(K) = FR(KSORT)
                FR(KSORT) = TEMPR
                TEMPI = FI(K)
                FI(K) = FI(KSORT)
                FI(KSORT) = TEMPI
            ENDIF
  300 CONTINUE
C  Normalize
        NORM = 1.0/SQRT(FLOAT(NF))
        DO 400 K = 1, NF
            FR(K) = NORM*FR(K)
            FI(K) = NORM*FI(K)
  400 CONTINUE
        RETURN
        END
        INTEGER FUNCTION BITREV(J, NU)
C  J has NU bits; reverse into BITREV
        INTEGER J,NU,JT,I,JD
        JT = J
        BITREV = 0
        DO 100 I = 1,NU
          JD = JT/2
          BITREV = BITREV*2+JT-2*JD
          JT = JD
  100 CONTINUE
        RETURN
        END
```

These programs for the FFT will enable you to make fast, efficient discrete Fourier transforms of the EEG voltage data in order to extract the Fourier amplitudes. The same programs, merely with IT switched in sign, may be used with the Fourier amplitudes as input to predict the EEG voltages. It's probably worth your effort to understand the FFT algorithm in A5.8 and the FFT program in this section just to save work when programming the EEG voltage prediction.

Predicting voltages from Fourier amplitudes

As the next step in the Fourier analysis, as described in the program structure at the end of L9.1, we insert the Fourier coefficients just calculated into the defining relation (L9.1) and make predictions, $V_p(t)$, for values of t from 0 to 1 sec. Note that we may also use $V_p(t)$ to interpolate voltages to times when data are not available. Extrapolation of voltages to points outside the range $0 < t < 1$ sec. will merely reproduce the voltages predicted within this time interval. Such a behavior is unlikely for an EEG. However, a Fourier analysis of a physiological quantity with an observation interval of 1 day might be extrapolated over times of more than a day to search for biological rhythms with a period of about a day (circadian rhythms).

▶ *Exercise L9.2.3*   (Recomposing EEGs)
This exercise consists of adding a program segment to reconstruct the voltage at chosen values of t from the a_k and b_k computed in the preceding exercise.
(a) Use the conventional Fourier transform algorithm to evaluate (L9.1) directly. This is readily accomplished by using a pair of nested loops in which the outer loop is over the time index j and the inner loop is over the frequency index k. The predicted voltages should be stored in an array for use in L9.3. The output should be a table labeled by time, and/or time index, and by the predicted voltage $V_p(t)$ in μvolts.
(b) Alternatively, apply the fast Fourier transform algorithm to reconstruct the voltage $V(t)$. The FFT procedure can be used almost the same as before, except for the sign change of imaginary part implied by the sign change in the exponents between (A5.7) and (A5.8). Therefore, the amount of extra coding required for the use of the FFT in this mode is very small.
(c) Display the original data, $V(t)$, the predicted voltages, $V_p(t)$, and $10[V(t) - V_p(t)]$, all on the same graph (with different symbols) as a function of t. Comment on the quality of the reproduction, especially why there should be any discrepancy at all. If you have the display procedure from L4, it would be a convenient option to get your program to make the display. □

This completes the programming required for the spectral analysis of the EEG. Section L9.3 examines the sources of error in greater detail. As a diversion for engineers, we suggest the following exercise:

► *Exercise* L9.2.4   (Testing Wiener-Khinchin)

(*a*) Test how well the Wiener-Khinchin theorem (see Exercise A5.2.2) is satisfied by your numerical results. To do this, add a program segment to compute the sum of the squares of the Fourier amplitudes and the sum of the squares of the input voltages, $V(t)$. According to the theorem, they should be equal (within the round-off error of the computations). If they differ significantly more than this explain why. (*b*) To convince yourself that the difference is not due to round-off errors, use the predicted voltages $V_p(t)$ instead of $V(t)$. Now the agreement should be much closer. (If the agreement isn't better, check your coding!) ☐

In this section of the computing laboratory we have implemented the discrete Fourier transform and its inverse. This can be done essentially without error. In the next section we investigate two common error sources in Fourier analyses of data.

L9.3 {3}   The Nyquist criterion and noise

Uncertainties in the Fourier coefficients arise from two major sources of error: First, the finite number of data points, here $N \leqslant 32$, and second, random noise in the voltage data. In this section we examine these error sources, mainly by numerical computations.

One estimator of the error in our Fourier analysis is the chi-squared, χ^2, here defined by

$$\chi^2 = \Sigma \, [V(t_j) - V_p(t_j)]^2, \qquad (L9.4)$$

with the sum running over j from 1 to N. The standard deviation between observed and predicted values, S, is given by

$$S = \sqrt{[\chi^2/(N - 1)]}, \qquad (L9.5)$$

assuming that $N > 1$. In these formulas equal errors of 1 unit in each observation are assumed. The values of χ^2 or S provide an overall estimate of the reliability of the fit. According to the least-squares principle discussed in sections A4.7 and A5.1, the Fourier coefficients minimize χ^2.

► *Exercise* L9.3.1   To convince yourself that you can't do better than the Fourier coefficient prescription, add a program option that allows you to change any of the a_k or b_k between their computation and their use in predicting the voltages. Verify that any changes you make in the coefficients make the value of χ^2 increase. ☐

Now that we have χ^2 as an error estimator, we can investigate two major error sources in Fourier analyses applied to data.

The Nyquist criterion

The errors in the Fourier analysis that arise from the finite number of data points, also called the *sampling errors,* are related to Nyquist's criterion, discussed in A5.2. Nyquist suggested that if there is not at least one sampling point in each period of the highest frequency Fourier component required, then the Fourier coefficients for this frequency will be unreliable. This criterion is tested in the following exercise.

▶ *Exercise L9.3.2*   (Testing the Nyquist criterion)

(*a*) Write a program segment to compute the standard deviation, S, between any two samples contained in arrays of the same length. Use this program segment to calculate and output S for the original EEG data and the predicted voltage obtained using the full 32 data points.

(*b*) Reanalyze the EEG using only every other data point, so that NDATA is now 16, rather than 32. Now the frequencies from 17 to 32 Hz will have zero amplitude. Recompute the standard deviation between the original data (32 points) and the predictions obtained by using the Fourier coefficients produced with the reduced number of data points. By what fraction has the standard deviation increased over the value obtained in (*a*)?

Continue halving the number of data points but using the same 1-second range of t values, recalculating the coefficients and recomputing S until you are dissatisfied with the quality of the prediction. Does your criterion for a good fit agree with Nyquist's? For this part of the analysis it should make no difference, except for efficiency, whether the CFT or FFT is used. □

Now that we have gained some understanding of the error source from sampling, we consider the effects of noise superimposed on the signal.

Effects of noise

The concept of noise has been discussed in the context of computational error in A4.2 and in the context of random numbers in L6.1 and L6.2. We now use it in the context of signal processing. Here a noise is a random signal superimposed on the signal that we are measuring. What constitutes a signal and what is considered noise is dependent on what one is trying to measure.

For example, the noise signals in the infrared detectors at the Bell Laboratories at Holmdel, New York, in the early 1960s were considered bothersome noise by physicists Penzias and Wilson, until they realized that they were detecting remnants of cosmic black-body radiation from early stages in the evolution of the universe. Thus, one man's noise is another man's Nobel prize. An interesting account of Penzias and Wilson's research and its outgrowths is given in the *Scientific American* article by Muller.

In an EEG analysis the noise is likely to originate from movement of the patient, from loose electrical connections, and from voltages generated by electrical equipment nearby, such as motors and computers. One usual criterion for noise is that it vary randomly with time. Its associated frequency range will therefore appear to be about the sampling frequency. Experience with pseudo-random numbers from L6 will therefore be helpful when investigating the properties and consequences of noise.

▶ *Exercise L9.3.3*   (Analyzing noise effects)
Include an option in your program to allow input of the RMS value of the noise signal to be superimposed on your original data. Then a program segment should modify the data by adding a random voltage to it at each time interval. To make the voltage random in sign as well as magnitude begin with pseudo-random numbers in the range 0 to 1, multiply them by 2, then subtract 1, according to the formulas developed in L6.1.

Repeat the Fourier analysis, the prediction of the voltages, and the calculation of the standard deviations as in Exercise L9.3.1. If you plot S as a function of the RMS noise level input, you can extrapolate back to zero noise input in order to infer the average noise level in the original data. It is also interesting to plot the Fourier coefficients as a function of the input noise in order to see which frequencies are most affected by noise. □

The computations in this section relate to the techniques of Fourier analysis. The last section of this computing laboratory introduces another technique for identifying the major frequency components in an oscillatory signal.

L9.4 {3}   Autocorrelation analysis of the EEG

The autocorrelation analysis method is often used as an alternative to, or in conjunction with, the Fourier analysis. Its use in interpreting EEGs is described in the book by Ledley, and in Chapter 9 of Hobbie's book. Autocorrelation analysis considers how a signal correlates with itself ("auto") as a function of the lag time between elements of the signal. The correlation will be strong if the lag time is a multiple of the period of components having the largest amplitudes. A typical autocorrelation for EEG data is shown in Figure L9.4.1.

We derive the properties of the autocorrelation function as follows. The autocorrelation function that we use, $C(\tau)$, is defined by

$$C(\tau) = \int dt \ V(t)V(t + \tau), \tag{L9.6}$$

in which the integration is over the range of times for which data exist. (It is also common to define the autocorrelation by summations over data points, a definition that is well adapted to use of the FFT and that is more suitable than the present one if extensive computations are to be made.)

The autocorrelation C is a function of the parameter τ, called the *lag time*.

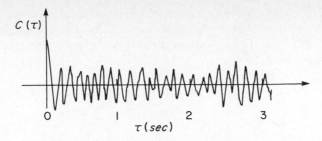

Figure L9.4.1   Autocorrelation as a function of lag time τ for a normal EEG. The *a* rhythm is predominant.

For each value of τ, the integral is to be repeated in order to produce a new value of C. From our definition (L9.6), a number of practical properties of $C(\tau)$ can be deduced, as in the following.

Properties of autocorrelations

Autocorrelations, defined according to (L9.6), may be computed for any set of data $V(t)$. Here we emphasize how the properties of $C(\tau)$ relate to the Fourier amplitudes of $V(t)$. Given the Fourier expansion of the signal

$$V(t) = \Sigma \left[a_k \cos (2\pi kt) + b_k \sin (2\pi kt) \right], \tag{L9.7}$$

the right-hand side can be inserted in the definition (L9.6) of $C(\tau)$. If summation and integration are interchanged, and trigonometric functions of $(t + \tau)$ are expanded in terms of those of t and τ, and if the integration is over one cycle of the fundamental frequency $k = 1$, then one finds straightforwardly that the autocorrelation can be expressed in terms of the Fourier coefficients as

$$C(\tau) = \Sigma \left[a_k^2 + b_k^2 \right] \cos (2\pi k\tau), \tag{L9.8}$$

in which the summation is over all frequencies k for which the Fourier coefficients are nonzero.

▶ *Exercise* L9.4.1   Show explicitly the steps leading from (L9.6) to (L9.8) for the autocorrelation $C(\tau)$. ☐

▶ *Exercise* L9.4.2   Show that the Wiener-Khinchin theorem (that the total power is the same whether computed in the time or in the frequency representations) is obtained from the autocorrelation formula (L9.8) by setting $\tau = 0$, and comparing the result with the definition (L9.6). ☐

Formula (L9.8) shows that $C(\tau)$ has a Fourier cosine expansion, all of whose coefficients are positive. Those frequencies with the largest magnitudes of a_k and b_k will dominate the oscillations of the autocorrelation as τ varies. Thus, if there is a predominant frequency k_m in the Fourier spectrum, then $C(\tau)$ will tend to oscillate at this frequency. Further connections between autocorrelations and Fourier expansions are presented in Brillouin's book on science and information theory.

Noise and autocorrelations

Suppose that we have a voltage signal $V_s(t)$ and some noise $r(t)$. The observed voltage is then

$$V(t) = V_s(t) + r(t),\tag{L9.9}$$

with r randomly distributed in time. Since the autocorrelation defined by (L9.6) involves integrating a product, expansion of V into two terms will produce four terms in the autocorrelation. First, there will be the autocorrelation of V_s, then the correlations of V_s with r and of r with V_s, and finally the autocorrelation of the noise r itself. However, the noise should be uncorrelated with the signal and with itself, so that the autocorrelation is that due to the signal alone.

▶ *Exercise* L9.4.3
(a) Go through the mathematical steps in the argument just presented to show that if the noise is uncorrelated with the signal, then

$$C(\tau) = C_s(\tau) + \int dt\ r(t)r(t + \tau).\tag{L9.10}$$

(b) Show that $C(\tau) = C_s(\tau)$ except at $\tau = 0$, and that this result still holds at $\tau = 0$ in the approximation that the root-mean-square (RMS) noise is negligible compared with the signal RMS value. □

Note that even in the absence of noise the values of the autocorrelation function computed from the definition may differ from those computed from the Fourier expansion (L9.7) because of the upper limit on k in the expansion and because of approximations made in computing the integral in (L9.6).

▶ *Exercise* L9.4.4   (Autocorrelation analysis program)
(a) Compute and output the autocorrelation function $C(\tau)$ of the EEG voltage, using the definition (L9.6) directly. Assume that the data are periodic with period 1 sec, so that if $t + \tau$ is negative then values of V at $t + \tau - 1$ are used. This is called *wraparound*.
The integral may be approximated by the trapezoid rule (see A4.4 and L5.2). If

wraparound is assumed and the trapezoidal formula is used, then the autocorrelation may be approximated by a sum

$$C(\tau) \approx \Sigma V_j V_R, \tag{L9.11}$$

in which the sum is over all the data points j = 1 to j = N, t is the integer nearest to $\tau/\Delta t$, and R = mod $(j + t, N)$. The *mod* function in the formula for R takes care of the wraparound. Note also that the step-size factor h from the trapezoid formula has been omitted since the overall magnitude of $C(\tau)$ is not of interest. Indeed, one sometimes uses the normalized autocorrelation, $C(\tau)/C(0)$, which is scaled so that it is unity at zero lag time, τ = 0.

The program segment for this option will consist of a pair of nested loops, the outer loop over an index labeling τ, say $N/2$ values, and the inner loop over a time index j running from 1 to N.

(*b*) Display the autocorrelation as a function of τ. A hand-drawn graph or use of the plotting procedure from L4 is suitable. Is there a periodic behavior of C? If so, is the associated frequency that for which the Fourier amplitudes obtained in L9.2 are largest in magnitude?

(*c*) An alternative to computing the autocorrelation directly in terms of the definition (L9.6) is to use the expression (L9.8) in terms of the Fourier amplitudes. Write a program segment to do this. Its structure will be similar to that in (*b*), except that the inner loop over the time index j is replaced by a loop over frequency index k. How closely do the results from (*b*) and (*c*) agree? Which involves the less total computing effort to obtain $C(\tau)$?

(*d*) To examine the effect of noise on the autocorrelation function, add noise (as pseudo-random numbers) to the original EEG data and recompute the autocorrelation pattern. Is the autocorrelation insensitive to noise, as (L9.10) claims? □

This exercise on autocorrelations shows that this quantity can be very useful in extracting the main features of a periodic pattern. The procedure is so straightforward that specially programmed microprocessors, called *correlators* and *signal averagers*, are used to evaluate and display estimates of $C(\tau)$, even while the signals are being accumulated. To quote a recent advertisement: "Whether your application involves turbulence and acoustic analysis, laser light scattering correlation, time-of-flight determinations, or similar types of signal processing, we have the instrument to do the job efficiently and economically."

In this computing laboratory you will have explored many techniques for signal analysis and for practical calculation of Fourier expansions. You will find the following references helpful in extending your knowledge and practical experience in applied Fourier analysis.

References on Fourier analysis and the EEG

Boore, D. M., "The Motion of the Ground in Earthquakes," *Scientific American*, December 1977, p.68.

Brazier, M. A. B., "The Analysis of Brain Waves," *Scientific American*, June 1962, p.142.

Brigham, E. O., *The Fast Fourier Transform*, Prentice-Hall, Englewood Cliffs, N.J., 1974.

Brillouin, L., *Science and Information Theory*, Academic Press, New York, 1956.

Cameron, J. R. and J. G. Skofronick, *Medical Physics*, Wiley, New York, 1978.

Cooper, J. W., *Introduction to Pascal for Scientists*, Wiley, 1981.

Crichton, M., *The Terminal Man*, Bantam Books, New York, 1973.

Cromwell, L., F. J. Weibell and E. A. Pfeiffer, *Biomedical Instrumentation and Measurements*, Prentice-Hall, Englewood Cliffs, 1980.

Hasler, A. D., A. T. Scholz, and R. M. Horrall, "Olfactory Imprinting and Homing in Salmon," *American Scientist* **66**, 347 (1978).

Hobbie, R. K., *Intermediate Physics for Medicine and Biology*, Wiley, New York, 1978.

Ledley, R. S., *Use of Computers in Biology and Medicine*, McGraw-Hill, New York, 1972.

Merrill, J. R., *Using Computers in Physics*, Houghton Mifflin, Boston, 1976.

Muller, R. A., "The Cosmic Background Radiation and the New Aether Drift," *Scientific American*, May 1978, p.64.

Otnes, R. K. and L. Enochson, *Applied Time Series Analysis*, Wiley, New York, 1978.

Regan, D., "Electrical Responses Evoked from the Human Brain," *Scientific American*, December 1979, p.134.

Robinson, E. A., and S. Treitel, *Geophysical Signal Analysis*, Prentice-Hall, Englewood Cliffs, N.J., 1980.

Rose, J., "Cataclysmic Computing," *Datamation*, April 1981, p.168.

Ryan, J., *The Adolescence of P-1*, Ace Books, New York, 1979.

Schmidt, S. A., "Fourier Analysis and Synthesis with a Pocket Calculator," *American Journal of Physics*, **45**, 79 (1977).

Spehlmann, R., *EEG Primer*, Elsevier, New York, 1981.

Stanley, W. D., and S. J. Peterson, "Fast Fourier Transforms on Your Home Computer," *Byte*, December 1978, p.14.

Von Neumann, J., *The Computer and the Brain*, Yale University Press, New Haven, 1958.

Walter, W. G., "The Electrical Activity of the Brain," *Scientific American*, June 1954, p.54.

L10 ANALYSIS OF RESONANCE LINE WIDTHS

This computing laboratory is on the analysis of resonance line widths, with application to the determination and maintenance of time standards by the use of atomic clocks. The laboratory brings together concepts and results from many sections of this book; resonances from A7.4, Fourier integral transforms from A5.7, transform computations by the numerical integration methods in A4.4 and L5, and straight-line least-squares fitting from A4.7 and L8.

In L10.1 we have a brief introduction to atomic clocks and their relation to resonance phenomena in the microscopic realm. In section L10.2 we particularize to the study of the inversion resonance in ammonia, which occurs in the microwave frequency region of the electromagnetic spectrum. This resonance is observed as an enhanced absorption of microwave power by the ammonia gas in a narrow frequency range about the resonance frequency. Since the precision of an atomic time standard depends on the line width γ of the resonance used, one needs to determine this width precisely. However, γ of a Lorentzian-shaped resonance has a very nonlinear relation to the power absorption spectrum, so that linear-least-squares methods of extracting γ are not directly applicable.

We show in L10.3 how to transform the absorption spectrum by use of Fourier integral transforms and logarithms to a form in which γ appears linearly. Thereby a linear-least-squares fit can be made to obtain the resonance width. These data transformations incur errors, and some sources of these are analyzed in L10.4. The chapter concludes with references on resonances and atomic clocks.

L10.1 {2} A brief introduction to atomic clocks

How can atoms or molecules, and resonance phenomena involving them, be used as clocks? To answer this, it is most instructive to consider why the daily motion of the Earth around its axis and the yearly motion of this axis around the sun were previously used as time standards. A macroscopic system such as Earth is expected to maintain a steady motion, relatively unaffected by perturbations, at least over periods of individual human experience. A disadvantage of this time scale is that it is difficult to subdivide to time intervals much less than human experience, for example to the millisecond range.

Atomic motions as clocks

Another way to obtain periodic motions that are insensitive to external perturbations, and which can therefore be used as time standards, is to use the microscopic realm of atoms and molecules. One could use the periods of electrons in their atomic orbits, or the resonance frequencies of atoms in molecules, as in the ammonia atomic clock, or the transition frequencies between different angular momentum (spin) states of electrons in atoms, as in the cesium atomic clock. The basic periods, which are in the 10^{-11} second range or smaller, can be transferred by optical, electronic, and mechanical means to longer time scales.

One advantage of atomic clocks over the use of Earth's rotation as a time standard is in transportability. Atomic clocks that had a precision in the nanosecond (10^{-9} sec) range were used by Hafele to fly around the Earth in jet aircraft in order to test the time dilatation predictions of special and general relativity theories. (This experiment is described in the article by Hafele and Keating referenced at the end of this chapter.) In the search for extraterrestrial intelligence (SETI), reviewed in the book edited by Goldsmith, it is much easier to communicate the time standard based on a (presumably) universal atomic phenomenon rather than on the rather local properties of a minor planet of an ordinary star.

The early development of atomic clocks is surveyed in a *Scientific American* article by Lyons, and the use of microwaves is discussed in an article by Davis. There is an extensive article under "Atomic Clocks" in the *McGraw-Hill Encyclopedia of Science and Technology*. The research journal for measurement standards, *Metrologia*, has a special issue in 1977 devoted to articles on the accuracy of time standards. Recent developments are indicated in the article by Hellwig, Evenson, and Wineland, and in that by Muyall, Daams, and Boulanger. The accuracy and precision of atomic clocks has allowed minute variations in the periods of the Earth's rotation to be determined and used for a variety of geophysical and astronometric purposes. These topics are presented in the book edited by McCarthy and Pilkington that is referenced at the end of this chapter.

Resonances and clocks

How can a resonance frequency, as discussed in A7.4, be used as a clock? If we assume that the resonance frequency is truly constant, but a clock indicates that it is varying, then we must assume that the clock period is varying. If in a time interval T the resonance frequency appears to have changed by an amount Δf from the nominal frequency f_0, then, for relatively small changes, the comparison clock time is wrong by an amount ΔT, where

$$\Delta T/T = -\Delta f/f_0. \qquad (L10.1)$$

▶ *Exercise* L10.1.1   Derive this formula for the relation between time and frequency changes by using differential calculus and assuming that the derivatives can be approximated by finite differences. □

We expect that Δf should be proportional to the full width at half maximum (FWHM), γ, of the resonance. Therefore, the precision (reproducibility error) of the standard clock is ultimately limited by the sharpness of the resonance line. The accuracy (absolute error) of the standard clock is limited by the insensitivity of the resonance frequency, f_0, and γ to environmental conditions. (This is one time in which insensitivity to the environment is desirable.)

The 1982 definition of 1 second is the duration of 9192631770 periods of the resonance radiation emitted between two energy levels of ^{133}Cs. As an indication of the precision of such atomic clocks, it was reported in the magazine *Physics Today* in November 1976 that the atomic clock at the U.S. National Bureau of Standards has a reproducibility of 0.85 parts in 10^{13}. The measurements also implied that the international atomic second, maintained by the International Time Bureau in Paris, was too short by about 11 parts in 10^{13}. (This revelation makes the Universe about a day older than previously believed by scientists. However, as described in the book by Weinberg, most of the general dynamics of the Universe were determined by its evolution in the first three minutes, this loss of time should therefore not affect our present disposition, except for those inclined to astrology.)

L10.2 {2}   The inversion resonance in ammonia

One of the first molecular vibrations considered for use as an atomic clock was ammonia gas, NH_3. In NH_3 the H atoms form an equilateral triangle, as is required by their equivalence. The N atom occupies a mean position that is not in the plane of the hydrogen atoms, as shown in Figure L10.2.1. The vibrations are approximately harmonic, as our analysis will reveal. A more detailed analysis of the ammonia microwave spectrum using quantum mechanics is presented in Chapter 12 of the book by Townes and Schawlow. Both of these American physicists were awarded Nobel prizes for research

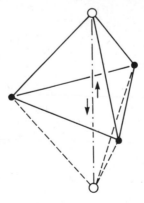

Figure L10.2.1   Configuration of the NH_3 molecule, showing the N atom (open circle) at its mean stable position above the triangular base of the H atoms (black dots). The dash-dot center line indicates the motion of the N near resonance.

related to microwaves: Townes was Nobel laureate in 1964 for microwave investigations, and Schawlow was honored in 1981 for his development of laser spectroscopy.

The ammonia inversion resonance data that we use were reported by Townes in a paper in the *Physical Review* in 1946. He found a resonance at a frequency f_0 = 23.870 GHz (1 GHz = 10^9 cycle/sec). Thus, since the microwaves are electromagnetic waves propagating at the speed of light, c = 3.00×10^{10} cm/sec., the wavelength at resonance is λ_0 = c/f_0 = 1.20 cm.

Microwave absorption experiments

In the microwave region the electromagnetic wavelengths are comparable to those used in acoustic experiments. However, a 1.2-cm sound wave in air has a frequency of 27,000 Hz, which, although well above audible frequencies for human beings (except Clark Kent), is still miniscule compared with that for the ammonia resonance. Therefore the techniques of using standing waves in laboratory size apparatus can be used straightforwardly. A sketch of a waveguide arrangement for microwave absorption experiments is shown in Figure L10.2.2. More detail is given in the article by Davis and in the book by Townes and Schawlow. The length of the resonance cavity should be several times the resonance wavelength, 1.2 cm for the ammonia inversion resonance. Near the resonance frequency a relatively large amount of power will be absorbed by the gas molecules in order to supply the kinetic energy of vibration of the resonant oscillations.

The dependence of the absorbed power, P, on frequency, f, is calculated in A7.4 for a harmonic system with damping proportional to velocity. It is given by the Lorentzian function

$$P(f) = P_0\gamma/\{[2\pi(f - f_0)f_0)]^2 + [\gamma/2]^2\}. \tag{L10.2}$$

Here we have included the conversion from oscillation frequency f (in Hz) to angular frequency ω (in radian/sec). The data are usually expressed with f as

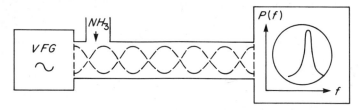

Figure L10.2.2   Schematic of microwave absorption apparatus. Variable frequency generator (VFG), waveguide cavity, and power absorption meter are indicated. The wavelength, 1.2 cm, is just that for the NH_3 inversion resonance.

the variable, but theoretical analyses are usually simpler with ω as variable. Note that at resonance $P(f_0) = 4P_0/\gamma$, rather than P_0. This choice is made for later convenience in the formulas. The data that we use are scaled to arbitrary units of power, so that the absolute value of P_0 will not be important in our analysis.

Since the precision of locating the resonance frequency f_0 is limited by its FWHM of γ, our goal will be to determine a best-fit value for γ, given microwave absorption data near the resonance frequency. Unfortunately, the relation between the data $P(f)$ and γ is highly nonlinear, as (L10.2) shows. Therefore, the usual straightforward methods of linear least squares, as covered in A4.7 and L8, cannot be applied directly to determine γ. Rather, we transform the data, by Fourier integral and logarithmic transformations, to produce a linear relation that allows γ to be extracted. The details of these transformations are presented in L10.3.

Data for the ammonia inversion resonance

Data on the inversion resonance in NH_3 are given in the article by Townes in the *Physical Review*. He found that the power $P(f)$ absorbed by NH_3 gas at low pressure depends on frequency as given in Table L10.2.1.

Table L10.2.1 NH_3 inversion resonance data

f	P	f	P
23.76	0.3	23.88	3.3
23.78	0.6	23.90	1.7
23.80	1.0	23.92	1.0
23.82	1.4	23.94	0.5
23.84	2.2	23.96	0.2
23.86	4.5		

Outside the range of frequencies given here the power absorbed from the microwaves by the ammonia gas may be assumed to be zero. The resonance frequency value $f_0 = 23.870$ GHz should be used. One way of estimating this value from the data is by a cubic-spline analysis, as outlined in Exercise L7.2.2 (d).

L10.3 {2}   Fourier-transform analysis of a resonance

In this section we first derive the Fourier cosine transformation of a Lorentzian-shaped resonance and show that this has an exponential decay with time. Therefore the logarithm of the transform will have a linear dependence on time. We show that the slope of this line is $-\gamma/2$. Then we discuss the interpretation of the transform and its relation to the complementarity principle in quantum mechanics. A program outline for the resonance analysis and several exercises that can be developed using the program are then presented.

Derivation of the transform relation

The Fourier integral transform of a Lorentzian function can be easily derived by the method of contour integration in the complex plane, as described in Wylie's book. However, we can dispense with such power if we are ingenious in this Wiley book.

We will use the Fourier cosine transform, the real part of the Fourier integral transform in section A5.7. That is, we compute

$$p(t) = \int d\omega \, P(f) \cos [(\omega - \omega_0)t], \qquad (L10.3)$$

in which the integral is over positive frequencies. Relative to the definition in A5.7, we have also introduced a phase difference ω_0, merely to simplify the subsequent analysis. In this equation t is a time variable, measured in time units reciprocal to those of the frequency f.

We compute the Fourier integral (L10.3) as follows. Noting that t appears in the integral only in the cosine function, we can differentiate both sides of the equation with respect to t twice and obtain

$$D_t^2 p(t) = -\int d\omega \, P(f)(\omega - \omega_0)^2 \cos [(\omega - \omega_0)t], \qquad (L10.4)$$

with the integral over positive frequencies. Now we use the Lorentzian expression (L10.2) explicitly, with the abbreviation $w = \omega - \omega_0$. On manipulating the integrand, one finds directly that

$$D_t^2 p(t) = (\gamma/2)^2 p(t) - P_0 \gamma \int dw \, \cos (wt), \qquad (L10.5)$$

with the integral now ranging from $-\omega_0$ to ∞. If the resonance frequency is many times γ above zero, then the effect of the integral over the cosine term is relatively negligible, because the oscillating integrand contributes nothing on the average. There thus results the differential equation for the Fourier cosine transform (L10.3) of the Lorentzian function

$$D_t^2 p(t) = (\gamma/2)^2 p(t). \qquad (L10.6)$$

▶ *Exercise* L10.3.1   Derive in detail differential equation (L10.6) for the Fourier transform of the resonance, using the steps just indicated. □

The solution of the differential equation (L10.6), studied in A7.3, is generally

$$p(t) = p_- \exp(-\gamma t/2) + p_+ \exp(\gamma t/2). \tag{L10.7}$$

If we are to calculate the Fourier transform only for times $t > 0$, then we must choose $p_+ = 0$ to keep $p(t)$ finite for large t.

▶ *Exercise* L10.3.2   Evaluate p_- in (L10.7) by noting that $p_- = p(0) = \int dw\, P(w)$, where the integration is from $-\omega_0$ to ∞. Show that if $\omega_0/\gamma \gg 1$, then the integral is just $\gamma/2\pi$. (To perform the integral, the transformation $2w/\gamma = \tan \theta$ is useful.) □

At last we have the complete expression for the Fourier integral cosine transformation of the Lorentzian

$$p(t) = 2\pi P_0 \exp(-\gamma t/2). \tag{L10.8}$$

The last step is to take the natural logarithm of both sides to obtain the linear relation

$$\ln [p(t)] = \ln [2\pi P_0] - (\gamma/2)t. \tag{L10.9}$$

Thus, a plot of $\ln [p(t)]$ against t should be a straight line with slope $-\gamma/2$.

Interpreting the Fourier transform

What is the interpretation of all this mathematics? Is it merely conjuring to produce a convenient form for linear- least-squares analysis? The exponential decay relation (L10.8) shows that the Fourier integral transform of the Lorentzian resonance $p(t)$ decays with a *relaxation time* $\tau = 2/\gamma$, which is the time required for the signal intensity to decay by a factor $1/e$. We interpret this as follows. If the source power is removed, then the forced vibrations decay in a characteristic time τ. For the ammonia inversion resonance, the N atoms return to their equilibrium motion outside the plane of the H atoms in a time inversely proportional to the damping constant γ.

For ammonia, $\gamma \approx 2\pi \times 0.06$ Grad/sec $= 0.37$ Grad/sec, as you can see by sketching the data from Table L10.2.1. Therefore, the relaxation time $\tau \approx 6$ nsec. (1 nanosecond $= 10^{-9}$ sec.) An alternative to the frequency resonance scan method, as Townes measured it, would be to measure the decay with time of the power signal from an initially excited sample of NH_3 molecules. This method is used routinely in research laboratories, where timing at intervals of 10^{-9} sec or less is feasible.

We notice a complementarity between the frequency response of a system and its time behavior; If γ is small the resonance is very sharp as a function of frequency. In the time domain, signals die away very slowly if γ is small, because the relaxation time τ is correspondingly large; $\tau\gamma = 2$. This "complementarity principle" survives the transition to quantum mechanics. Together with its generalizations, it was used by Niels Bohr in his work on the interpretation of measurements in quantum mechanics.

Resonance analysis program structure

Here is a suggested structure, outlined in pseudocode, for your program to make a Fourier cosine integral transform $p(t)$ of the resonance data, then to make a linear-least-squares fit of ln $[p(t)]$ against t to extract the resonance line width γ.

    L10 Analysis of resonance line widths

    Data and options input
    Fourier cosine transforms
       Loop over times, t
          For each t integrate P(f) times cosine over frequency
             and store in array PCOS labelled by time index
       End loop over times

    Loop over times, converting PCOS to ln (PCOS)
    Plot PCOS against t to find linear segment (Option)
    Make least-squares fit of line to find slope $-\gamma/2$
    Predict P(f) against f; numerical and graphical output (Options)

    Go for more control data or options

    Program end.

The numerical integration to construct the Fourier transform could be done by using a program segment written in A4.4 and L5 (for trapezoid or Simpson formulas) or by using a library program available in your computer system. In making endpoint corrections for these integrations it might be worthwhile, especially for the trapezoid rule, to assume that there is an additional data point with value zero at each end. This provides a smooth transition to zero away from the resonance region. The graphical output is suitably handled by the printer plot given in L4.

The least-squares fit could be made using the programs written in A4.7 or L8, or a program obtained from a scientific subroutine package in your computer system.

Exercises on resonance analysis

The following two exercises provide extensive practice in the programming, graphing, and interpretation of the inversion resonance calculation for ammonia.

▶ *Exercise L*10.3.3   (Resonance line width program)

(*a*) Write, debug, and run a program for Fourier cosine transform analysis of Lorentzian resonance line widths, as just outlined. A useful test case for program debugging is to input $P(f)$ as an exact Lorentzian generated with known resonance frequency and FWHM, γ. Indeed, it would be a useful program option to have such data permanently in the program ready for use in testing subsequent modifications of it.

(*b*) The graphical output (by hand or using the plotting procedure from L4) should display the resonance data as distinct points at the data values, along with a smooth Lorentzian curve generated by your program at much closer steps of frequency than the data points.

(*c*) Assume that, with some sophistication, the center of a resonance line can be located to one-hundredth of its width. Then the magnitude of the limiting fractional accuracy $\Delta T/T$ for an atomic clock based on a resonance phenomenon would be $\Delta T/T = \Delta f/f_0$, which is $10^{-2}\gamma/f_0$. What is the fractional accuracy predicted from your analysis of the line width of the inversion resonance in NH_3? What other timekeeping devices have an accuracy of this order?

(*d*) Given the number of significant figures for the current standard clock, ^{133}Cs, quoted in L10.1, and assuming a Lorentzian for the frequency dependence, estimate the order of magnitude of the equivalent Δf for the Cs atomic clock.

(*e*) In our analysis we have not described how a best-fit value of the resonance frequency f_0 is arrived at. Suggest how this might be done, preferably within the scope of the Fourier transform analysis method. □

The comparison of forces in the microscopic realm with those in the macroscopic world is always of interest. For a final exercise on atomic clocks, work the following if you have time:

▶ *Exercise L*10.3.4   (Spring constants binding atoms in molecules)

(*a*) From Avogadro's number calculate the mass, m, of a nitrogen atom. Given the inversion-resonance frequency for ammonia, $f_0 = 23.87$ GHz, use the relation between the force constant, k, m, and f_0, $f_0 = \sqrt{(k/m)}$ from A7.4, to estimate the equivalent spring constant between N and H. Note that a reduced mass should be used, rather than the mass of the N atom alone. How to calculate this mass is not obvious; justify your choice.

(*b*) Given Young's modulus for iron, $Y = 2 \times 10^{11}$ dyne/cm^2, calculate the force constant k per pair of Fe atoms. Which bond is stronger, the N-H bond in NH_3, or the Fe-Fe bond in metallic iron?

(*c*) Discuss the nature of the "springs" that bind atoms together in molecules and in metals. □

L10.4 {3}   Error analysis for the Fourier transform

From your computations of the Fourier transform of the NH_3 inversion resonance, you will have discovered that for both small and large values of t the
transform $p(t)$ differs significantly from a straight line. Why is this so? I have
thought of two major sources of error in this analysis, perhaps you can find
others. The error sources discussed here are common to a wide class of data;
they are called *finite-range-of-data* errors and *finite-step-size* errors. In the context
of the discrete Fourier transform, such errors are considered in L9.3. We consider each error source briefly.

Finite-range-of-data errors

The specification for calculating the Fourier transforms given in (L10.5)
requires that the integral be performed over the complete frequency range.
Since the data $P(f)$ peak at the resonance frequency f_0 (for which there may
not necessarily be a data point) and decay fairly rapidly (depending on γ), it is
not obligatory to have data much beyond several line widths on each side of
the resonance peak. However, our data in Table L10.1 are truncated while
still about 5% of the peak value.

When t is small, the truncation (finite-range-of-data) errors are most severe
because the cosine term in the Fourier transform integral (L10.5) varies
negligibly from unity over the range of variation of the data. The situation for
the Fourier integral is illustrated in Figure L10.4.1. Therefore we are essentially computing the average value of the power $P(f)$ at low values of t. Since

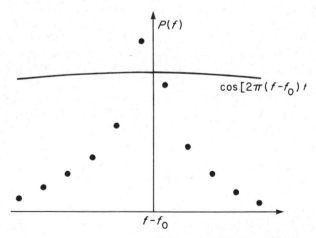

Figure L10.4.1   Truncated power spectrum of NH_3 resonance, and cosine function appearing in integral (L10.3),
showing the finite-range-of-data error in Fourier transform
calculations for small times.

the power absorbed is always positive, if we have a finite range of data we will underestimate the integral. Thus, as you should see from your graph of ln $[p(t)]$ against t, the values for small times dip below those extrapolated from a straight-line fit to the middle range of times.

Finite-step-size errors

Another source of error in the Fourier integral transform of data comes from the discreteness of the data used in an integration which implies infinitesimal spacing of the values. Perhaps we should have used the discrete Fourier transform, as in the EEG analysis in L9. However, there are relatively few functions whose discrete Fourier transform can be derived analytically. Therefore, in the present analysis we would have lost the advantage of knowing the relations, (L10.3) and (L10.9), between the parameters of interest (here γ) and the Fourier transform.

When t is very large, such that $t \gg 1/(\omega_{max} - \omega_0)$, where ω_{max} is the largest frequency used in the data, the cosine term oscillates very rapidly in sign as t varies. Since only discrete values of ω appear in the data, with discrete data the apparent values of the integral (L10.3) will fluctuate rapidly and thereby give an unreliable and fluctuating estimate of $p(t)$. This behavior is indicated in Figure L10.4.2. The effects should be evident in your Fourier transforms for large time. They are essentially the same as those discussed as the Nyquist criterion analyzed in L9.3.

Therefore, in analyses that require the computation of integral transforms on data, finite-step-size errors must be expected. The optimum solution is to

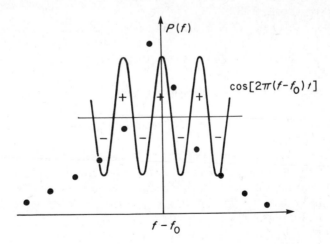

Figure L10.4.2   Finite step size of the power spectrum data and a rapidly varying cosine function for large times, leading to unreliable estimates of the Fourier transform.

reduce the step size in the independent variable, but this requires more time, sometimes more equipment, and nearly always more money.

▶ *Exercise* L10.4.1   (Error analysis for integral transforms)

(*a*) Investigate finite-range-of-data error effects in integral transforms by removing from the ammonia inversion resonance spectrum one or more data points from each end. Does the fit become even more nonlinear for small times?

(*b*) Demonstrate finite-step-size errors in integral transforms by removing alternate data points from the NH_3 inversion resonance data, then repeating the analysis. Are the large values of t most affected by this change? □

In this computing laboratory we have investigated how to transform data to a more convenient representation, applied particularly to a Lorentzian resonance in a molecular system, NH_3, in order to extract the resonance width. We also learned how to determine the Fourier integral transform of a resonance line shape analytically and numerically, the latter process requiring care in the selection of data because of the sources of error.

There are several topical text books on the use of transform techniques in applied science. Their use in chemistry is discussed in Griffith's book, in digital signal processing there is a book by Oppenheim and Schafer, the text on signal analysis by Papoulis has particular emphasis on the theory of Fourier transforms, and the monograph by Blass and Halsey describes practical methods of deconvoluting absorption spectra.

References on resonances and atomic clocks

Blass, W. E., and G. W. Halsey, *Deconvolution of Absorption Spectra*, Academic Press, New York, 1981.

Davis, M. H., "Radio Waves and Matter," *Scientific American*, September 1948, p.16.

Goldsmith, D. (ed.), *The Quest for Extraterrestrial Life*, University Science Books, Mill Valley, Calif., 1980.

Griffiths, P. R., *Transform Techniques in Chemistry*, Plenum Press, New York, 1978.

Hafele, J. C., and R. E. Keating, "Around-the-World Atomic Clocks," *Science*, **177**, 166 (1972).

Hellwig, H., K. M. Evenson, and D. J. Wineland, "Time, Frequency, and Physical Measurement," *Physics Today*, December 1978, p.23.

Lyons, H., "Atomic Clocks," *Scientific American*, February 1957, p.71.

McCarthy, D. D., and J. D. H. Pilkington (eds.), *Time and the Earth's Rotation*, D. Reidel, Dordrecht, The Netherlands, 1979.

McGraw-Hill Encyclopedia of Science and Technology, "Atomic Clocks," McGraw-Hill, New York, 1982.

Metrologia, **13**, No.3 (1977), a special issue on the accuracy of time standards.

Muyall, A. G., H. Daams and J.-S. Boulanger, "Design, Construction, and Performance of the NRC CsV1 Primary Cesium Clocks," *Metrologia*, **17**, 123 (1981).

Oppenheim, A. V., and C. W. Schafer, *Digital Signal Processing*, Prentice-Hall, Englewood Cliffs, N.J., 1976.

Papoulis, A., *Digital Signal Processing*, McGraw-Hill, New York, 1977.

Townes, C. H., "The Ammonia Spectrum and Line Shapes Near 1.25-cm Wave-Length," *Physical Review*, **70**, 665 (1946).

Townes, C. H., and A. L. Schawlow, *Microwave Spectroscopy*, Dover, New York, 1975.

Weinberg, S., *The First Three Minutes*, Basic Books, New York, 1977.

Wylie, C. R., *Advanced Engineering Mathematics*, McGraw-Hill, New York, 1975.

L11   SPACE-VEHICLE ORBITS AND TRAJECTORIES

Computers, from micros to mainframes, are essential in a wide range of tasks relating to space vehicles such as space shuttles, orbiting spacelabs, satellites, and interplanetary probes. In this laboratory the uses of the computer that we investigate in detail are the calculation of satellite orbits from formulas, the numerical solution of trajectory equations for motion under gravity, computer file storage of data relating to space vehicles, and the display of satellite orbits.

The major sections of this laboratory and their connections with other chapters are as follows: Section L11.1 gives an overview of scientific uses of space vehicles and the importance of computers in these applications. It also reviews how satellites are used in determining Earth's resources and in telecommunications, including long-range computer networks. In section L11.2 we calculate satellite orbits according to the analytic formulas derived in A8. Also in this section, satellite trajectories are calculated by the numerical methods for solving differential equations that were developed in A6.6 and A7.6. Output of the orbit or trajectory data to a computer file, to enable later retrieval and use, is also discussed in this section. Display of satellite orbits or trajectories is suggested in L11.3, using the computer graphics techniques and programs developed in L4. We conclude this laboratory with extensive references on space vehicles and their applications.

L11.1   {1}   Space vehicles, satellites, and computers

Many of the applications of computers to space science are well known through the attention they receive when a malfunction or inconsistent output halt the launch of a space vehicle. The news media emphasize the use of on-board computers in space missions. However, more computer resources are used before a space vehicle is launched and during the acquisition and interpretation of data than are used on the vehicle during flight. The article by Stakem has a good overview of the applications of computers in space missions. In this section we review uses of space vehicles, communications satellites, and data on Earth satellites.

Uses of space vehicles

We consider briefly four types of space vehicles and their uses; space shuttles, space stations, Earth satellites, and interplanetary probes.

The *space shuttle* is a space vehicle launched by rocket to altitudes of 150 km and higher into orbits of small eccentricity. The shuttle has sufficient on-board power to allow it to be decelerated from a stable orbit then to enter Earth's atmosphere and land as a glider. It is designed such that most of its components can be reused and so that the refurbishment time is of the order of weeks or less. There is an extensive article entitled "Space Shuttle" in the *McGraw-Hill Encyclopedia of Science and Technology*. A major use of space shuttles is for the deploying, repairing, and retrieval of space satellites. In other applications, an interesting article on space shuttles for terrestrial remote sensing is that by Taranik and Settle.

A *space station* is a large space vehicle in semipermanent orbit, with living and experimental facilities for occupation by several scientists for extended periods. A discussion of the scientific results from the Skylab space station is given in the book by Kent, Stuhlinger, and Wu. The long-term prospects for the use of space stations are discussed in Napolitano's book, including such applications as materials processing in zero gravity. The article, "Space Station," in the McGraw-Hill encyclopedia details many applications.

Earth satellites are the most common space vehicles. Their uses range from the investigation of Earth resources (as discussed in the article by Joels), geological studies (as detailed in the articles by Muehlberger and Wilmarth and by Flinn), to the use of satellites as signal relay stations, the so-called communications satellites that are discussed in more detail following.

These three types of space vehicles are, at the time of writing, Earth-bound. However, *interplanetary probes* are space vehicles accelerated to velocities exceeding the escape velocity from Earth orbit so that they can be directed to other planets. Problem [A8.5] examines some of the dynamics of optimal launching of a space vehicle for interstellar exploration. Interplanetary probes must be maneuvred very intricately if the gravitational attraction of each planet visited is used to best advantage. The planning for such interplanetary missions therefore requires very detailed numerical computations of the planetary and space vehicle motions. The technical details of navigation between the planets are given in a *Scientific American* article by Melbourne and in books by Ehricke and by Thomson. The results from such interplanetary probes are presented in books by Murray and Burgess and by French and in a series of articles on Voyager 2 in *Science*.

Communications satellites

A communications satellite is designed to receive signals from Earth-based stations, to amplify the signals, then to retransmit them to Earth. Modern communications satellites are put into an orbit such that they appear stationary when observed from Earth. The situation is indicated in Figure L11.1.1. For a geosynchronous satellite the period, T, is 24 hours. This requirement suggests some interesting exercises on design considerations for geosynchronous communications satellites:

▶ *Exercise* L11.1.1   (Geosynchronous satellites)
(a) Show that the altitude above Earth's surface of a geosynchronous satellite (one whose period is the same as that of Earth) is 35,860 km (22,282 mile).

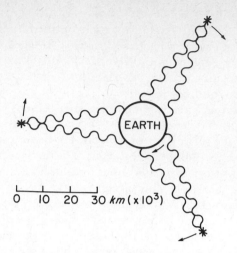

Figure L11.1.1 Geosynchronous satellite arrangement shown to scale. The three satellites indicated give global coverage, except for the polar regions.

(*b*) Show that three geosynchronous satellites would be sufficient to broadcast to all of Earth except the polar regions at more than 81° latitude.

(*c*) Why must a geosynchronous satellite be situated above the equator?

(*d*) Signals to and from communications satellites travel at essentially the speed of light, 300,000 km/sec. Show that the time delay in a to-and-from trip to a geosynchronous satellite varies from 239 msec directly below the satellite to 279 msec in the fringe viewing area. (Ignore any delay in the satellite's transmitter regenerating the signal.) □

Geosynchronous satellites are very useful as communications satellites because the Earth stations do not need expensive tracking equipment, they can be used to intercommunicate between several Earth stations (each satellite is in view from 42% of Earth's surface), and Doppler shifts in signal frequencies arising from relative motion between source and receiver are negligible. However, the signal power is reduced greatly compared with a low-altitude asynchronous satellite, because of the inverse-square-law attenuation. With a powerful transmitter on the satellite, and a sensitive receiver on Earth, the use of communications satellites (with attendant receiving dishes on the ground) is becoming very common.

Communications satellite systems are discussed extensively in the 1978 book by Martin and in the article "Communications Satellites" in the *McGraw-Hill Encyclopedia of Science and Technology*. The more general considerations of data communications, including those between computer systems, are covered in the book by Fitzgerald and Eason.

Data on Earth satellites

For comparison with the numerical calculations in the next section, it is interesting to have data on Earth satellites. The orbits of most space shuttles and space stations are nearly circular and are at low altitude, typically 150 km. However, to put some adventure into our space calculations, it is more interesting to consider eccentric satellites, which are primarily those launched in the 1960s. Table L11.1.1 gives data on some representative satellites. Apogee, r_a, and perigee, r_p, altitudes are in km, and the periods, T, are in minutes.

Extensive data on Earth satellites are given in the article "Satellite" in the *McGraw-Hill Encyclopedia of Science and Technology*. You may use these data in the next section when calculating the orbits and trajectories of satellites.

L11.2 {2}   Numerical methods for orbits and trajectories

In this section we develop the methods for the numerical computation of satellite motion in a bound orbit in Earth's gravitational attraction. The calculation of orbits, using the formulas developed in A8.3, is straightforward once the orbit parameters have been determined in terms of easily observed quantities. The numerical integration of trajectory equations to give the time development of the radial and angular motions is more involved. It requires use of numerical methods for second-order differential equations, such as those developed in A7.6. We try the second and fourth Euler approximations.

In this laboratory we consider only stable orbits and are not concerned with how the orbit is established. This vital part of any space mission, from launch to transfer into the final orbit, is calculated using extensions of the numerical methods derived here. The technical details are given, for example, in the books by Ehricke and by Thomson.

Table L11.1.1   Data on eccentric satellites

Satellite	r_a	r_p	T
Sputnik I	900	230	96
Explorer II	2540	350	114
Vanguard	3950	650	134
Sputnik III	2790	190	116
Telstar II	9900	960	215

From observables to orbital parameters

Here we derive and assemble the formulas needed to convert from observables to the quantities needed to compute orbits and trajectories. We take as observables for the orbit the radius at perigee, r_p, and the radius at apogee, r_a. Figure L11.2.1 shows the observables and orbital parameters. Appropriate units for distance are kilometers (km) and for time, minutes (min). For satellites, the altitude above Earth's surface is usually quoted, as in Table L11.2.1, so that the radius of Earth, r_E = 6367 km, needs to be added to the altitudes. Note that the period is that referred to a fixed point at the center of Earth. This differs from that observed from a point on Earth because of its diurnal rotation.

For the orbit, we have from A8.3 on the geometry of ellipses that

$$r_0 = 2r_a r_p/(r_a + r_p), \qquad (L11.1)$$

and the eccentricity

$$\varepsilon = (r_a - r_p)/(r_a + r_p). \qquad (L11.2)$$

It is useful for the trajectory calculations and the display of the orbit to have the semi-major axis a and semi-minor axis b available. From A8.3 we have for an ellipse

$$a = (r_a + r_p)/2, \qquad b = \sqrt{(r_a r_p)}. \qquad (L11.3)$$

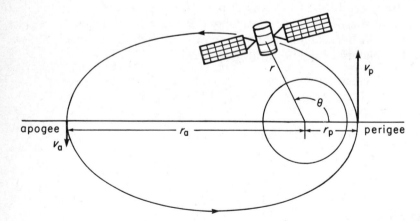

Figure L11.2.1   Observables and orbital parameters for an Earth satellite. The variables are discussed in the text. The eccentricity of this orbit is 0.8.

▶ *Exercise* L11.2.1

(a) Check out Kepler's third law, (A8.42), for the satellites given in Table L11.1.1 by calculating T^2/a^3 for each of them.

(b) What is your best estimate and standard deviation for *GM* from these values?

(c) Derive the relation $GM = gr_E^2$, where g is measured at the Earth's radius, $r_E = 6367$ km. Is the value of g derived from these relationships consistent with that which you experience daily? □

Also from A8.3, we have the formula for the orbital radius, r, as a function of the angle, θ (in radians), from perigee

$$r(\theta) = r_0/(1 + \varepsilon \cos \theta). \qquad (L11.4)$$

With these preliminaries, we are in a good position to calculate satellite orbits analytically.

▶ *Exercise* L11.2.2   (Program for orbits)

(a) Use the orbit equation (L11.4) to write a program segment that generates an array of r as a function of θ. The necessary inputs are r_a and r_p, which you may obtain by using Table L11.1.1. The loop index of the array should be that for controlling θ. The program should make a loop over angles from zero to 2π in steps of about $\pi/20$ rad $= 9°$. If you wish, the angle scale may be in degrees, in which case loop over angles from zero to 360° by 10° steps; however, make sure that the angle is in the appropriate measure (usually radian) before the cosine is taken. The use of the full angular range is not necessary, because of the symmetry of the trip from perigee to apogee with respect to the trip from apogee to perigee; however, it is useful for the display section in L11.3. To check the orbit program, try a circular orbit ($r_a = r_p$), and also verify that for any eccentricity the orbit repeats for every change of θ by 2π.

(b) Make an output option to produce a labeled table (with units at the head of the table) of r against θ. This output will be useful for checking the trajectory results from the numerical integration. □

Now that you have calculated satellite orbits, you should be ready for the next step in space, the calculation of satellite trajectories.

Numerical integration for trajectories

The satellite trajectory describes its path as a function of time. Therefore it contains more information than the orbit, which is like a time exposure photograph of the trajectory. For the same reason, extra parameters relating to time or forces must be provided. Choose one or the other, but not both, of the period T or *GM*. If you use both, experience shows that they may not be consistent with the orbital parameters. This will give trouble when you try to calculate an accurate trajectory.

Because we are investigating the *time* dependence of the motion, we need to calculate a time-dependent quantity such as a velocity. Our choice for establishing the velocity scale is the tangential velocity component at perigee, v_p. Since perigee is an extreme point of the motion, the radial component of the velocity there is zero, as at apogee. We can calculate the areal velocity at perigee and, since it is constant (according to Kepler's second law), we may equate it to the average value over one period, T. Thus

$$r_p v_p / 2 = \pi a b / T, \tag{L11.5}$$

in which on the right side the formula for the area of the orbital ellipse has been used. Solving (L11.5) for the tangential component of the velocity at perigee, v_p, gives

$$v_p = 2\pi a b / (T r_p). \tag{L11.6}$$

In calculating the radial acceleration of the space vehicle we also need the constant of the gravitational force GM. (If you have a good vehicle you may not need GM.)

▶ *Exercise* L11.2.3   Show that if you do not know the universal gravitational constant G and the mass of Earth M, you can obtain their product from the relation

$$GM = g r_E^2, \tag{L11.7}$$

where g is the gravitational acceleration measured at Earth's surface, a distance r_E from its center. □

The differential equations to be solved for the radius are given by (A8.23) as

$$D_t^2 r = [(r_p v_p)^2]/r^3 - [GM]/r^2 \equiv AR(r) \tag{L11.8}$$

for the radial acceleration. The differential equation for the angular velocity is given by (A8.22) as

$$D_t \theta = [r_p v_p]/r^2 \equiv VTH(r). \tag{L11.9}$$

For efficiency, the quantities in brackets, [ ], in the preceding two equations should each be treated as a unit that is precomputed and saved before the time evolution of the trajectory is calculated. It is even clearer in the programming to define the right-hand sides of equations (L11.8) and (L11.9) by functions, $AR(r)$ and $VTH(r)$, respectively, as indicated.

We investigate two of the Euler approximations discussed in A7.6 to integrate the differential equations for the motion of the satellite. We begin with the *second Euler approximation* for a second-order equation

$$v_{n+1} \approx v_n + a_n \Delta t, \qquad (\text{L11.10})$$

$$x_{n+1} \approx x_n + (v_n + v_{n+1})\Delta t/2, \qquad (\text{L11.11})$$

which is given as (A7.61).

▶ *Exercise* L11.2.4   (Integration by second Euler approximation)
(*a*) Transcribe the notation from A7.6 to that appropriate to the trajectory radial equation to show that in the second Euler approximation the pair of equations for advancing the radial motion is

$$VR_{n+1} \approx VR_n + AR_n \Delta t, \qquad (\text{L11.12})$$

where *VR* is the *radial* component of velocity, and

$$r_{n+1} \approx r_n + (VR_n + VR_{n+1})\Delta t/2. \qquad (\text{L11.13})$$

(*b*) Show that the second Euler approximation for advancing the angular motion is

$$\theta_{n+1} \approx \theta_n + (VTH(r_n) + VTH(r_{n+1}))\Delta t/2. \qquad (\text{L11.14})$$

☐

Note that the equations should be used in the sequence given, because of the feed-through of the velocity and radius predictions.

Alternatively, we may use the *fourth Euler approximation*, given as (A7.63),

$$v_{n+1} \approx v_n + a_n \Delta t, \qquad (\text{L11.15})$$

$$x_{n+1} \approx x_n + v_{n+1}\Delta t. \qquad (\text{L11.16})$$

▶ *Exercise* L11.2.5   (Integration by fourth Euler approximation)
Transcribe notations for the fourth Euler approximation in A7.6 to produce the iteration formulas for numerical integration of the radial motion

$$VR_{n+1} \approx VR_n + AR_n \Delta t, \qquad (\text{L11.17})$$

$$r_{n+1} \approx r_n + VR_{n+1}\Delta t, \qquad (\text{L11.18})$$

and the iteration formulas for the angular motion

$$\theta(r_{n+1}) \approx \theta_n + VTH_{n+1}\Delta t. \qquad (\text{L11.19})$$

☐

These formulas should be used in the indicated sequence because the estimate of r at different n values feeds through them.

Before the integration can begin, we must specify the initial conditions. If we
choose the time origin to coincide with perigee, which we chose as $\theta = 0$,
then we have the following initial conditions: At $t = 0$ the radial component
of velocity $VR_0 = 0$, for the initial radius $r_0 = r_p$, and for the initial angle
$\theta_0 = 0$.

Structure of the satellite program

From the detailed algorithms for the orbit and trajectory calculations as given
earlier, we can outline the structure of the satellite program. The program segments on data files and on displays are straightforward to code, but are quite
dependent on the computer system used. They are also indicated in the
following pseudocode outline:

> L11 Space-vehicle orbits and trajectories
>
>> Data and options input
>> Convert from observables to orbital parameters
>
> L11.2 Numerical methods for orbits and trajectories
>
>> Analytic calculation for orbits; Loop over angles θ
>> For each θ, calculate $r(\theta)$ into array with θ index
>> Save array of radii for data files or display (Options)
>> Output table of θ and radii (Option)
>
>> Numerical integration for trajectories
>> Choose second or fourth Euler approximation (Option)
>> Loop over times, t
>> For each time, advance solution of differential equation
>>   to next time interval, storing radius and θ into
>>   arrays with time index
>> End loop over times
>
>> Data files for space-vehicle data (Option)
>> Output or input the data file used to generate orbit and
>>   trajectory (Options: Output/input and orbit/trajectory)
>
> L11.3 Display of satellite orbits (Option)
>> Find range of data, scale it, and locate origin
>> Initialize plotting array to ground character
>> Locate plotting points in array with figure character
>> Output plotting array

Go for more data or options

Program end.

All the ingredients for the step-by-step integration of the radial and angular equations of satellite motion are assembled. It is now time for you to use the recipes given and burn off. (However, this is not a cook-book laboratory.)

▶ *Exercise* L11.2.6   (Program for trajectories)
(*a*) Write a small program segment to calculate the trajectory parameters according to equations (L11.1) through (L11.3). Output these parameters as part of the program verification and for education, if not enlightenment.
(*b*) Add a program segment to initialize radial velocity, radius and angle. Output these values. Do not confuse the radial velocity VR with the tangential velocity at perigee v_p. In particular, at perigee VR is *zero*. The output values from (*a*) and (*b*) can be checked for the case of a circular orbit ($r_a = r_p$) by a simple hand calculation.
(*c*) Add to your program a loop over time t, which is labeled by the index n. Code the Euler approximations, given by equations (L11.12) through (L11.14) for the second Euler method and by (L11.17) through (L11.19) for the fourth Euler method. The choice of which algorithm is used should be a user option; do not mix together the two algorithms, unless you are curious about the outcome! The values of the radius and angle should be stored in arrays indexed by the loop control variable. Run your program with test data. A suitably accurate time step is $\Delta t \approx 0.1\,T$. Unless you have all parameters and formulas correct, it is very unlikely that you will get anything resembling an elliptical orbit. Here are some methods of checking your program.
(*d*) To check the numerical integration, run a circular orbit, which has $r_a = r_p$. The velocity components should not change with time; the radial velocity, VR, should be always zero, and the tangential velocity at perigee and always, $v_p = 2\pi r_p/T$, as follows from the distance/time formula for unaccelerated motion.

If the circular orbit result holds, try running an eccentric orbit (not in the athletic sense) and output the radial component of velocity, VR, at each time step Δt. The flight from apogee to perigee should have negative radial velocities, and these should be nearly equal in magnitude to those for their corresponding angles on the flight from perigee to apogee. Also, as follows from the previous discussion, the radial velocity should be zero at both apogee and perigee.

As a final check, you should be able to fly for more than one orbit and nearly overlap the earlier trajectory. However, all these checks are conditional on the adequacy of the numerical approximations made in the Euler algorithms. These are tested in (*e*) and (*f*).
(*e*) Investigate the effects on the numerical accuracy of the trajectory calculation of varying the time step Δt. For small eccentricities, say less than 0.2, $\Delta t/T \approx 0.1$ should be adequate. Don't be tempted to keep on reducing Δt if this doesn't hold. You might just be covering up programming bugs, because all the Euler formulas (unless grossly miscoded) give exact results as $\Delta t \to 0$.
(*f*) Compare the second and fourth Euler algorithms by running fairly eccentric

orbits, say $\varepsilon \approx 0.8$ (which has $b/a = 0.6$), and seeing how well each of the two Euler algorithms trace out the orbits. For this it's also a good idea to have the analytic orbit results for the same space-vehicle parameters. According to the article by Cromer, the fourth Euler algorithm should be especially suited for the type of bounded motion that we are studying here. Is this what you find? □

These algorithms for space-vehicle trajectories are also suitable for programmable pocket calculators. A program similar to our second Euler algorithm is described in the book by Henrici. The force equations can also be integrated directly in Cartesian coordinates using a program for a pocket calculator, as described in Eisberg's book.

Files for space-vehicle data

An important aspect of computing in applied science is the storage and retrieval of data in computer-readable form. In a mission as involved as launching, maneuvering, and recording data from a space vehicle, millions of bits of data are generated, and they must be retrieved at very high rates. Therefore, the use of data files, both in on-board and ground-control computers, is necessary.

The devices that have been used for computer data storage are mostly magnetic recording devices. For example, in early computers and on very small computer systems, the recording device was a magnetic tape. More commonly, from microcomputers to macrocomputers, magnetic disks of various kinds are used. There is a wide range of storage capacities, from the 5.25-inch floppy diskette with a capacity of 137 Kbyte on the Apple microcomputer that I'm using as a computer terminal, to the 10-inch hard-disk drive with a capacity of about 50 Mbyte on the IBM-3081 macrocomputer that is storing the text of this book. Two interesting references on disk-storage technology are the *Scientific American* article by White and the entry "Computer Storage Technology" in the *McGraw-Hill Encyclopedia of Science and Technology*. Computer data-base organization is well described in the 1975 book by Martin, which is suitable for a beginner in this area.

If the programming language that you are using is convenient for handling data files, then it would be an interesting exercise for you to record the orbit or trajectory of a satellite after you have calculated it. Practice retrieving the data and printing it in a later computer session. One difficulty with such exercises, and why I am unusually vague in suggesting this assignment, is that the handling of data files is strongly dependent on the programming language used. Within each language, file handling is dependent on the dialect and on the computer system used. In Pascal, the Esperanto of programming languages, you will find structures for handling records and files as part of the language. However, the way these are actually used for a given computer system varies among computers. Fortran, the Latin of the scientific programming languages, is not very well adapted to handling of data files, and implementation of file handling is usually system dependent.

In the options for data files in your program there will be one pitfall; how do you ensure that no attempt is made to read from a file before data have been written to it? The computer system will usually check for this, but it will probably also terminate your program. One possibility is to code a program segment that is used only to initiate files and to set a flag as the first record of the file. This flag should be changed as soon as your program writes data into the file. The flag value could always therefore be checked to see whether data were present or not.

L11.3 {2}   Display of satellite orbits

The visualization of scientific data provides a very efficient means of communicating information. For example, if you have completed the satellite orbit or trajectory program segments in L11.2 you will probably have several printouts of r against θ, and of r and θ against time. But, since the actual data represent motion in space and time, isn't it easier to show someone what is happening by means of a picture than by means of a table of numbers? This does not deny the usefulness of the numbers, which are certainly vital for calculating.

In this section it is suggested that you make a display of space-vehicle orbits or trajectories using the methods discussed in the introduction to computer graphics in L4. The simplest and most available method is the printer-plot method for which sample programs in Pascal and Fortran and Pascal are given in L4.2. However, the spatial resolution will be quite coarse, so if you have access to one of the graphics devices discussed in L4.3 you should probably make use of it. Whatever method you use, the structure of the printer plotting procedure given in L4.1 will probably be helpful.

Geometry of the orbit display

The geometry of the display involves two main considerations; converting from polar to Cartesian coordinates and determining origins and scales in x and y directions. The geometry is straightforwardly handled, with the conversion between the satellite polar coordinates $[r, \theta]$ and its (x, y) coordinates being given by

$$x = r \cos \theta, \quad y = r \sin \theta. \tag{L11.20}$$

Note that the coordinate center is the force center, the center of the Earth, rather than the origin of the plotting space. Therefore, either you must displace the origin correctly or it is simpler just to let the routine that finds the range of the Cartesian coordinates (MINMAX in L4.2) do this work for you. This latter method is suitable for orbits, as long as you are willing for the scales to be different for each choice of major axes. Otherwise, you should force some appropriate scales yourself. For trajectories, the latter is probably the best method, because if the orbit is at all inaccurate the scales will vary from one calculation to the next.

In plotting the elliptical orbits there are two problems that you should antici-pate. The first problem is the relative scaling between x and y directions. Since most display devices do not have the same scales in the two dimensions, you will need to compensate for this in your own display. It is especially tricky because an ellipse plotted with differing scales projects into an ellipse, as derived in Exercise A8.3.5. Suggestions for checking the scaling are given in Exercise L4.2.1 (*b*).

Illusory ellipses

The second problem in displaying elliptical orbits relates to human perception of projections. We are very accustomed to viewing circular objects and images at a slant angle and interpreting them as circles, whereas they appear on our retinas as ellipses. When an ellipse is presented to us we usually try to see it as a circle. (This perceptual bias is not only psychological but also cultural; see many of the articles in the *Scientific American* reprint collection on image, object, and illusion, edited by Held.)

The best way around the perceptual problem of interpreting ellipses as cir-cles is to provide a reference circle. for example, the figure of Earth within the space-vehicle orbit. This requires computation of the Earth's Cartesian coordi-nates

$$x_E = r_E \cos \theta, \quad y_E = r_E \sin \theta. \tag{L11.21}$$

These can be computed in the same program loop that transforms the orbital coordinates. Also, this reference circle will help you in choosing the correct x and y display scales.

After completing the computing laboratory exercises in this and other chap-ters, you will be well prepared to explore the many exciting territories to be discovered while computing in applied science.

References on space vehicles and satellites

Cromer, A., "Stable Solutions Using the Euler Approximation," *American Journal of Physics*, **49**, 455 (1981).

Eisberg, R. M., *Applied Mathematical Analysis with Programmable Pocket Calcula-tors*, McGraw-Hill, New York, 1976, Ch.6.

Ehricke, K. A., *Space Flight*, Van Nostrand, New Jersey, 1960.

Fitzgerald, J., and T. S. Eason, *Fundamentals of Data Communication*, Wiley, Santa Barbara, 1978.

Flinn, E. A., "Application of Space Technology to Geodynamics," *Science*, **213**, 89 (1981).

French, B. M., *Mars: The Viking Discoveries*, NASA (1977).

Held, R. (Ed.), *Image, Object, and Illusion: Readings from Scientific American*, W. H. Freeman, San Francisco, 1974.

Henrici, P., *Computational Analysis with HP 25 Pocket Calculator*, Wiley, New York, 1977.

Joels, K., "Remote Sensing - Satellite Analyses of Earth Resources," Mercury, July-August 1977, p.13.

Kent, M. I., E. Stuhlinger, and S-T. Wu (Eds.), "Scientific Investigations on the Skylab Satellite," *Progress in Astronautics and Aeronautics*, **48**, MIT Press, Cambridge, 1976.

Martin, J., *Computer Data-Base Organization*, Prentice-Hall, Englewood Cliffs, N.J., 1975.

Martin, J., *Communications Satellite Systems*, Prentice-Hall, Englewood Cliffs, N.J., 1978.

McGraw-Hill Encyclopedia of Science and Technology, "Communications Satellites," "Computer Storage Technology," "Satellite," "Space Shuttle," "Space Station," McGraw-Hill, New York, 1982.

Melbourne, W. G., "Navigation Between the Planets," *Scientific American*, June 1976, p.58.

Muehlberger, W. R., and Wilmarth, V. R., "The Shuttle Era: A Challenge to the Earth Scientists," *American Scientist*, **65**, 152 (1977).

Murray, B. C., and E. Burgess, *Flight to Mercury*, Columbia University Press, New York, 1976. Ch.6.

Napolitano, L. G. (Ed.), *Space Stations Present and Future*, Pergamon Press, Oxford, 1976.

Stakem, P., "1 Step Forward - 3 Steps Backup: Computing in the U.S. Space Program," *Byte*, September 1981, p.112.

Taranik, J. V., and M. Settle, "Space Shuttle: A New Era in Terrestrial Remote Sensing," *Science*, **214**, 619 (1981).

Thomson, W. T., *Introduction to Space Dynamics*, Wiley, New York, 1961. Ch.4.

"Voyager 2," articles in *Science*, **215**, 499-587 (1982).

White, R. M., "Disk-Storage Technology," in *Scientific American*, August 1980, p.138.

INDEX

313